Also, by Bill Corsar

If Only These Walls Could Talk
Dancing to a Different Tune
Attraction of Opposites
Running Wilde

GONE
BUT NOT FORGOTTEN

BILL CORSAR

This is a work of fiction. The characters and events are figments of the author's mind.

Copyright © Bill Corsar
All rights reserved.

ISBN: 9798333481665
Independently Published

Cover and Typesetting by:
Creative Covers

oorwullie448@gmail.com

#	Chapter	Page
1	Anger Can Be Fatal	1
2	Grim Reapers	10
3	Abigail Watson	12
4	Jack Gibson	23
5	Rose Prentice	33
6	Megan Wilson	40
7	Arthur Morgan	46
8	Jacob Marshall	52
9	Richard Ainsworth	58
10	Simone Jessop	64
11	The Call	70
12	The Transplant Centre	76
13	To Sleep, Perchance to Dream	82
14	Sweet Dreams?	88
15	When Will It Start Working?	95
16	Preparing For Discharge	97
17	After Discharge	105
18	I Have A Dream	113
19	A Surprising Meeting	121
20	More Dreams?	127
21	Even More?	134
22	What Have We Learned?	137
23	It's Time We All Met	141
24	A Meeting Of Minds?	144
25	Interview Protocol	154
26	A Life In Eight Memories	158
27	Where To Start?	168
28	Fact Or Speculation?	175
29	Sophie's Biography	184
30	Face To Face Meeting	199
31	All Talk?	203
32	Inconclusive Conclusion	213
33	One Down	219
34	Coroner's Inquest	229
35	I Know Your Secret	238
36	An Unexpected Call	247
37	Charlotte And Arthur Meet	250

38	Getting Away With Murder?	257
39	All Clear?	264
40	Cut Worms Forgive The Scalpel	266
41	Life After A Death	274
Authors Note		287

1

Anger Can Be Fatal

"Bloody Roberto Caruso, I'll have you, you little dago wop shit," cursed Sophie.

Sophie Coleman was driving back along the M25 after a meeting with the loudly cursed Roberto Caruso, the manager of one of her company's hotels in Norfolk.

She had gone to the hotel to check why her business plan was working for all the other hotels in her region but not for this one. All of the others under her care had shown some uptake in revenue. The Angel Arms was managed by the accursed Bloody Roberto Caruso, and its revenues were declining.

The meeting had not gone well. Caruso blamed Sophie's plan. He was not impressed by her argument that all of the other hotels in her region were showing improvement. His hotel was different, and her plan was not working there. In an angry response, Sophie pointed out several areas where Signor Caruso had deviated from her plan.

He was adamant that he had interpreted the plan using his vast managerial experience gained in top-notch hotels in Italy. The fault was hers. Accusations flew back and forth with no resolution, and very late in the day, Sophie stormed out and drove home very angry.

What added to Sophie's anger was Caruso's attitude and language. He raised his voice and spoke at and over her while waving his arms around like a demented dago wop traffic cop. Not that Sophie voiced that particular opinion. She was careful not to give him any avenue of escape when she submitted her report on his failings to his superior.

"Who the hell does he think he is? I'll rip out his heart and feed it to him on a stick. He is for the chop in the morning if I have any say in the matter. Jumped up little shit, I'll show you who's boss."

Sophie's mood and anger matched the foul weather. It was a dark early February evening. It had been raining most of the day with occasional flurries of snow. Adding to the miserable driving conditions was the spray from speeding cars. Her windscreen wipers were working at full speed.

As she drove, she slapped the steering wheel and swore. Her anger had increased the longer Caruso spoke to her. It was not just what he said, but it was also the way he treated her. It reminded her of her dead husband, Zac.

He had started to treat her in the same dismissive manner not long after they got married. He belittled her and made her feel as if he was a superior male while she was a mere woman. It annoyed Sophie that Caruso's tone and attitude had reminded her of her husband.

"Fuck, fuck, bloody fuck," cursed an angry Sophie. As she cursed and swore, she slapped her steering wheel again with her right hand. It hurt, and she swore again. "Shit that hurt," she whined.

Sophie's vision was 20:20, but her angry mind was elsewhere as she drove. Her perfect vision was clouded by her rage. The meeting with Roberto Caruso had wound her up tight as a drum. It was a dangerous situation at the best of times on the motorway. Tonight, the conditions on the M25 were far from ideal for driving.

Sophie was on the inside lane of the motorway, following the traffic, but her lack of concentration meant she had ventured too close to the car in front. The danger of driving too close in bad weather was lost in the back of Sophie's mind somewhere behind the wall of hostility that was fixated on Roberto Caruso.

The drab grey asphalt of the M25 was hidden under a layer of slush and water that glistened red from the tail lights of the cars in front and brilliant white from the headlights of the vehicles on the opposite carriageway. The spray from the weight of speeding traffic on the road created a mist, making visibility even more difficult.

Suddenly, she saw the angry red, marbled glare of dozens of brake lights ahead of her, like a gigantic firework display distorted by the water streaming down her windscreen. In a moment of panic, she slammed on her brake pedal and turned the steering wheel to the left to avoid the vehicle in front. She was trying to make for the relative safety of the hard shoulder.

Tyre manufacturers spend time and money designing the tread on their product to be as safe as possible in wet conditions. It is said that good quality tyres can clear the equivalent of a bucket of water from a road every seven seconds. To do this efficiently, the vehicle wheels should be rotating.

Sophie's hard braking meant that her wheels were not rotating. As a result, a wedge of water was driven under the four tyres, lifting them away from the road and reducing the friction that allowed her to control the vehicle.

She suddenly realised her danger when she saw that she was heading for the rear end of the car in front. She grappled with light steering, which is common in aquaplaning. She did not hit the vehicle full-on but caught its passenger side back wing as she tried heading for the safety of the hard shoulder.

The first sound Sophie heard was the squeal of her tyres as she slid forward at speed. This was followed by the crunch of metal on metal. The crash created a rapid stop that sent a message to her airbag mechanism. There was a loud band as her airbags deployed. Sophie was restrained by her seatbelt, but her head was catapulted forward into the bag, which then deflated as she hit it as it was designed to do.

At this point in her crash, Sophie's chances of survival with some relatively minor injuries were high. What happened next radically changed that dynamic.

She almost breathed a sigh of relief just as a big truck slammed into the rear of her car. A hard rear-end crash in a vehicle creates whiplash. Sophie's head was catapulted back into her seat. Her head hit the seat headrest, which was not in an optimum position to be helpful in this situation. The impact of her head on the edge of the headrest knocked Sophie unconscious, so she did not see the carnage that ensued.

Her car was bulldozed off the motorway and across the hard shoulder by the heavy truck. As her nearside wheels hit the shallow muddy dip in the grass verge at the edge of the hard shoulder, her car tilted and flipped onto its passenger side. It eventually came to a stop on its side on the grass verge. She was now several feet past the edge of the hard shoulder, with her car underbody visible to the other cars on the motorway.

Sophie's limp body was suspended in her seat by her seatbelt. She was slumped against the central console with her head, arms and legs flopping to the side.

The carnage on the motorway created difficulty for the emergency services. The tailback behind the crash had grown long by the time they arrived. Some drivers had started to park on the hard shoulder. They had to be moved back onto the road to allow the emergency vehicles to attend the accident.

They only got to Sophie about half an hour after her crash. In that time, she had slipped in and out of consciousness several times.

Everyone in the chaos was looking out for themselves or their loved ones. Any help that could be offered was given to the nearest vehicle to their own. No one was around to hear Sophie's whimpered calls for help during her occasional bouts of consciousness.

The initial action of first responders when they arrived at Sophie's car was an assessment by the medical team. She was alive but mostly unresponsive to light and sound. They then moved aside for the extraction team from the fire service. The extraction team got her car into a position where she could safely be moved. The team used a long backboard to slide Sophie out of the wreck. Another quick assessment of her critical medical needs was performed before she was rushed to Chase Farm Hospital, the nearest suitable A&E department.

During the ambulance trip, Sophie was still mostly unresponsive, with eyes occasionally flicking open. The medic in the back of the ambulance did what she could to manage Sophie. Her job was to give the A&E staff as favourable a chance as possible.

At the hospital, Sophie, an unconscious patient, had priority in the A&E.

Although unconscious, Sophie had shallow breathing, and there were signs that she had a basal skull fracture. She had clear fluid leaking from her nose and ears. The level of unconsciousness suggested to the doctors that there may be more extensive brain damage. A contrecoup injury of the brain was the obvious suspect.

A contrecoup brain injury is likely to occur when the head receives an impact and the soft floating brain rebounds in the opposite direction against the solid bone of the skull. A head thrown right might cause an injury to the left side of the brain. This is a fairly common injury in vehicle crashes. Sophie would be found to have several such injuries.

The attending medics wasted no time and immediately put out a call for the hospital duty crash team.

Their patient had some bruising on her chest from the seatbelt. There were some cuts on her face, arms, and legs and a penetrating wound to her eye. None of her veins or arteries appeared compromised. The only fluids leaking from her came from her nose and ears. Her body looked remarkably intact, given the seriousness of the crash.

The medical and technician teams worked tirelessly for several hours to identify the extent of Sophie's injuries in a vain attempt to keep her alive.

"Everybody stop," said Mr David Ratcliffe, the crash team leader. "We've done all we can for this lady, I'm going to call it. Anyone disagree?" There were no dissenters.

"Death occurred at one forty-four a.m. Thanks everyone."

"Shall I call the organ retrieval teams?" Asked Sheila Bedford, the sister in charge.

"Is she a registered donor."

"Yes, we found a card in her belongings, and the staff have checked the donor register. She's on that as well."

"Call them and keep her alive mechanically. She seems to be healthy apart from the fact that she is dead. She met the neurological criteria for brain death. The two teams might have a good harvest tonight. Does she have any next of kin details?"

"Her medical records list a Charlotte James in event of emergency. I called her an hour ago. I think she is in the waiting room. Do you want to speak to her?"

"I better had."

David Radcliffe cleaned himself up and went to the

waiting room, where many people were seated.

"Charlotte James?" He called.

Charlotte stood up and advanced towards him. "I'm Charlotte James. How is Sophie?"

"Would you like to come into my office please?"

In the office, he motioned Charlotte to the seat on the other side of his desk.

"Is Sophie OK?"

"I'm afraid that she died a few minutes ago."

"What happened?"

"She was involved in a road traffic accident earlier tonight and brought here unconscious. We discovered that she had suffered a significant brain injury which we could not fix. We did everything we could. I'm sorry for your loss."

"Oh," said a shocked Charlotte. "What do I do now?"

"How are you related to Sophie. I note your name is not the same as her."

"I'm her best friend. She has no relatives left alive. She asked me to be her next of kin in case something happened. I thought she meant illness, not death. I don't know what to do."

"I must ask you an important question."

"OK."

"Sophie is a registered organ donor. We would like to remove some organs for transplant. How do you feel about that?"

"I know she is a donor. We both registered at the same time. A close colleague at her work lost a young child two years ago. He died while waiting for an organ that never came. When she told me, we both thought it a good idea to register as donor."

"So, you're happy that we remove some of her organs for transplant?"

"I'm not happy she is dead, but yes, remove the organs you need. She would have wanted someone to benefit."

"Thank you, Charlotte. Do you want to view her body?"

"I don't know. What does it look like if she's been in a crash?"

"She looks quite well apart from a few fairly minor cuts and bruises."

"OK, I'll do it."

"I'll make sure she is ready and a nurse will come and take you to her. Once again, thanks."

Charlotte called her husband and told him the news.

"I'll need to go to her house and sort out her affairs. I'll do that later today. I've left an email with my work to let them know I won't be in later, and why."

"Do you know where everything is?"

"Yes. We spoke about it at length."

"Let me know if you need help."

"I'll be fine. Thanks love. I'll see you tonight when I'm done."

In the early hours of the morning of 12 February 2013, Sophie Coleman was about to donate eight of her organs to people she had never met or even know.

2

Grim Reapers

The call from Chase Farm's Accident and Emergency Department alerted two teams of organ harvesters working for the National Organ Retrieval Service in Bristol.

One team would come to work on the cardiothoracic area of Sophie, while the other would work on her abdomen. It was a challenging procedure as multiple individuals were working to access a small part of the body as rapidly as possible. The teams were in a race against time, not each other.

The first thing for the teams to assess was the suitability of the organs they might remove. If all organs are suitable for harvesting, the two teams must work out how to co-operate in such a small space.

Several professionals were competing for the best spot to harvest organs. The longer the delay in harvesting, the greater the chance that the organ would arrive at the transplant centre unusable.

A cardiothoracic harvester does not put their hand into the donor's chest to pull out the heart like some mad scientist in an old black and white B film. It's a skilful and delicate process designed to preserve as much of the heart's anatomy as possible. It is done this way to give the transplant surgeon the best chance of success.

After Sophie Coleman was pronounced dead, her lungs, liver, kidneys, pancreas, small bowel and the one remaining intact cornea were retrieved for donation. Each organ was examined and evaluated for quality. The information was communicated to the transplant coordinating team based in Bristol. It was their job to find the most suitable match for each organ.

The organs were packed and sent to the transplant hospital nearest to the homes of the patients considered most suitable to receive them.

Fortunately, these eight organs did not have long journeys. The longest journeys were to Addenbrooke's Hospital in Cambridge and the Churchill Hospital in Oxford. The rest went to hospitals in and around London.

At about three o'clock that same morning, eight people discovered that the death of an unknown person meant that they could receive a transplant. For three, it meant they had a chance to avoid an early, potentially painful death. As for the other five, it was a chance for a more normal life. On that particular night, the foresight and generosity of one person gave renewed hope to eight complete strangers.

3

Abigail Watson

After the war, Britain was a very drab place. Childhood in the 40s and early 50s would be described by those who lived through that time as life in black and white.

Abigail Gould was born in Norwich, Norfolk, just before the 1964 General Election that saw Harold Wilson's Labour government come to power.

By the time of her birth, much was changing. There was an optimism in British society at that time. This new, more confident mood coincided with the influx of more upbeat music, colourful fashion, technology, and holiday destinations for young people. All of these things were changing the social landscape of Britain from a drab black-and-white country into a more colourful one.

This was also a time when there were fewer cars, and children could play safely in the streets.

Abigail lived with her working-class parents in a typically small, two-up-two-down house in a row of identical terraced houses in the city. When she was three, she gained a sister with whom she would share a bedroom.

Despite the growth in labour-saving domestic appliances, the Gould family could not afford many of them. They had a larder instead of a fridge. It was a small room with a tiny window and a small air vent at the back of the house. As a result, food was usually stored in tins and bottles. Goods liable to perish had to be bought and consumed as soon as practical. In these houses, and at this time, use by dates consisted of touch, sight, taste and smell.

Abigail's father was employed on the railways. It was a time when the industry was going through significant change. The person responsible for the change was Dr Richard Beeching, a man admired or despised depending on one's stance on rail travel availability. It's worth noting that Beeching's boss, the transport minister Ernest Marples, was heavily involved with the company constructing the new M1 motorway, which was a direct rival to the railways. Beeching may have been the instrument of destruction, but Marples was the guiding hand.

Abigail's father earned just over £17 per week and paid £7 in rent for his council house. Her mother worked as a part-time cleaner to earn extra money for the family's non-essential expenses.

When Abigail started school, the first men were walking on the moon. Her first proper holiday was a short break during Easter weekend. Her family took her to the Butlins holiday camp at Skegness when she was six. The camp was about one hundred miles up the

English coast from her home in Norwich. Her favourite experience during this holiday was riding the monorail and chairlift at the camp.

At the time of Abigail's birth, almost ninety per cent of households owned or rented a TV set. Abigail's parents had rented their set from Radio Rentals.

The rise of popular music gave birth to the BBC television program 'Top of the Pops'. It was a half-hour show that compiled hits of the week performed by various artists. This program was launched the same year as Abigail's birth and ran for many years. This became one of Abigail's favourite television shows when she became interested in music and dance.

By the time of Abigail's early teens, the UK economy was about to be hit by conflict in the Middle East. The West's support for Israel during the Yom Kippur War angered the oil-producing Arab nations. They imposed an embargo on oil to the West, creating a financial crisis.

The price of petrol rocketed. Food prices were already high due to a global shortage. High petrol prices made transport even more expensive, adding to food costs.

The government response was to talk of using ration coupons left over from the Second World War. Abigail's parents knew all about rationing and coupons, having been children after the war. Rationing only ended ten years before Abigail was born. It was still fresh in her parent's memory.

As a result of high price inflation, unemployment rose, making life even more miserable for hundreds of thousands. Abigail's father was spared unemployment.

Her family's income did not rise to keep pace with inflation. Their tough life became even more difficult to bear.

Despite the creation of the welfare state, the problems it aimed to solve still persisted in society. Abigail's family situation serves as an illustration of the sluggish advancement made in this area.

Abigail was among the first generation of women who benefited from the introduction of the Equal Pay and Sex Discrimination Acts, as well as the rise of feminism. Women's struggle for the right to vote had been hard won. Despite these Acts, female equality was still some way off.

After World War Two, Britain saw a significant movement to improve the conditions of working-class children. These changes were aimed at improving their living conditions and education.

Although grammar schools were established to offer better opportunities for economically disadvantaged children, they were increasingly viewed as elitist, and very few students from underprivileged backgrounds attended them. To ensure educational equity, a new comprehensive school system was introduced to address the issue.

Abigail started her schooling at the Norwich Lower School, one of these new comprehensive schools. She moved to the Upper School when she began her secondary education.

She did not take full advantage of the education that was on offer. She saw schooling as something she was obligated to do and took no interest in any particular subject. Her financially struggling parents took little interest in her schooling and gave little or no encouragement to her studies.

Homework was a necessary chore that ate into her enjoyment of the popular music and culture of the day. Her favourite bands were the Beetles and the Kinks, but she moved on to some punk bands that rudely burst onto the music scene. Her parents were appalled by her experiments with Punk fashion and hairstyle.

She left school with minimal qualifications and an addiction to cigarettes. She was aware that smoking was not good for her health. She also knew that she was too young to buy cigarettes, but breaking the law was part of the excitement of smoking. She thought it was cool to smoke, as did many of her classmates and friends. They had egged her on to try it. At first, she choked but persisted and soon was addicted. Few people knew that not only was tobacco addictive, but tobacco companies added other addictive chemicals to their cigarettes to ensure repeat business. How they used their wealth to combat negative press is another story.

Local shopkeepers near schools broke the law and sold cigarettes in ones or twos to schoolchildren who had pocket money to buy them. The shop owners got more profit selling this way as they could demand any price they asked for their illicit product.

She was a teenager at the time of the violence wrought by the Angry Brigade and then the IRA. She was also alive during what became known as the 'Winter of

Discontent'. The violence and social struggles of the 1970s were not a time Abigail would look back on with fondness.

When she left school, she got a job as an assistant in the big Woolworth's store in Rampant House Terrace. This was the larger of the two stores the company had in Norwich. Abigail's poor educational efforts left her with limited employment choices.

While working behind the counter in Woolworths for a paltry wage, in the back of Abigail's mind was the idea that if she had worked harder at school, she might have got a better job than as a lowly paid assistant in Woolworths. A better job with a wage that allowed her to have more holidays in good hotels and warm places. She liked the idea of holidays abroad, but those were out of the question on her wages.

She did spend a lot of time at home watching television. She particularly liked the holiday shows and films showing exotic places worldwide. Abigail travelled to many parts of the world without leaving her sitting room.

Now that she was at work and had her own money, Abigail was soon smoking a twenty-pack of cigarettes a day, sometimes more. It never occurred to her to stop smoking and save her money for the holidays she desired.

She met Edward Gould at a dance in the Norwood Rooms just before it closed permanently in 1987. The two dated and eventually married in 1989. Like Abigail, Edward was a smoker and a dreamer. They often talked of places they'd like to visit if they had the money.

When they married, she moved to nearby Bury St Edmunds, where Edward was a plumber. The couple had one male child in 1991.

After moving to Bury St Edmonds with her husband, she got a job in the local flour mill. During her work in the mill, she was exposed to the dust of the flour mill. The combination of dust from the milling process and toxic smoke from her cigarettes soon began to affect her lungs. Her breathing gradually became more strained as her lung function decreased. She put her wheezing down to being a smoker and nothing more.

As a young woman, Abigail had been reasonably active without being an athlete. She played occasional recreational tennis in a local park during long summer days. In the winter, she played badminton with her girlfriends in a local club. The activities primarily involved socializing rather than engaging in sports.

As her lungs became more affected by her bad habits, her breathing deteriorated, and her weight increased, she became less active. This was not a conscious action; she would slow down or stop to rest when she ran out of breath.

She noticed her gradually increasing breathlessness. She continued to put her breathlessness down to smoking, being a bit overweight, and the fact that she was getting older.

It was her son who pointed out that her worsening health was probably caused by her tobacco habit. He encouraged both of his parents to stop.

"It'll help your health and save you money," he said.

"It's the only real enjoyment I get," she whined.

Her son, a non-smoker, continued to try to convince her to give up her habit. Abigail was addicted and not inclined to accept his advice. He warned her that her habit would kill her if she did not quit. His words were almost prophetic. Neither knew that her work in the flour mill only added to the smoking damage in her lungs.

She knew that smoking was hazardous to her health. The cigarette packages had the warning, and news media had been full of the dangers.

She did try to give up smoking several times, but the temptation was always close, in the form of her workmate's encouragement.

"Have a ciggy, Abby, one more won't kill you."

Little did Abigail know that even one more might. Every cigarette and every day at work in the flour dust was doing just that. She ignored all the other warning signs.

As her breathing deteriorated and routine exercise became more laboured, Abigail took a trip to her GP at the insistence of her son and husband. She did not think anything was wrong. She went because they badgered her into doing it.

"Anything to keep the peace, and you two off my back," she was heard to complain.

Following her initial physical examination, her GP instructed her to return for a spirometer test.

"Is there something wrong, Doctor?"

"It sounds as if there might be. I'll have a clearer

picture when I see the results of the test."

The following week, the practice nurse conducted the test. Abigail sat down while a small clip was placed on her nose. She was instructed to inhale deeply, filling her lungs, before exhaling forcefully into the spirometer mouthpiece. Meanwhile, the nurse encouraged her to continue exhaling even when Abigail thought her lungs were empty. The effort was painful. To make matters worse, the nurse repeated the test three times more. After the tests, Abigail felt like she had run a mile.

"What is that thing measuring?" Asked a breathless Abigail.

"It tells me the volume of air you can breathe out in a second, and the total amount of air you breathe out," said the nurse. "The doctor will compare those readings with normal readings for people of your age."

The nurse kept Abigail seated for fifteen minutes before she gave her patient a bronchodilator drug to inhale. The tests were then repeated to see if she had responded to the medication. Before she was to return for the results, she was also to attend the hospital for a chest x-ray.

Two weeks later, Abigail returned to the surgery to be given the results of her breath test and x-rays. Her sombre GP told her that the tests showed she had obstructive airway disease. He explained that her lungs could hold a typical amount of air. She could not breathe out as quickly as expected because she had a narrowing of her airways. This was what was causing her breathlessness.

There followed several other tests. They included a blood oxygen test, a peak flow test and an ECG.

After these tests, the GP convinced Abigail that her medical condition was not as trivial as she had wished. He warned her that it was a significant issue. He advised her to change her ways or face the consequences, which might be dire. She was to give up smoking and wear a mask in the mill. Unfortunately, she did not heed her doctor's advice and did neither regularly.

She took her prescribed, short-acting inhaler drugs, thinking this would be enough. It was not. The GP was forced to move her on to long-acting inhalers. He was becoming exasperated by Abigail's attitude to her health. He repeated his warnings about her deteriorating health and the risk that she might suffer an unpleasant death.

It took Abigail several more warnings from her GP and son to register that she had been stupid for not listening to either. The lightbulb moment came when the GP sent her to the clinic of Mr Adil Patel at the Addenbrooke's Hospital. She was so out of breath walking from the car park that she was placed on oxygen to aid her breathing.

Years of smoking and inhaling flour dust had damaged her airways to the extent that her physical activity became severely restricted. She had difficulty walking to the shops. They were only three hundred yards from her house. Her ability to work became impossible, and she was placed on disability benefits that included a blue badge. She was also prescribed oxygen therapy at home. Wearing the mask made Abigail feel old. She avoided looking at her reflection in the mirror while she wore it. She silently cursed herself for being stupid and ignoring advice.

The disease continued to progress to the point where all conventional forms of treatment had been exhausted, and a lung transplant was the only option if she was to avoid the unpleasant death implied by her GP.

Abigail was placed on the lung transplant register in 2012. Now, she could only wait and hope that someone died before she did to give her a chance at life. It was a selfish thought that was in keeping with her life to date.

When Sophie Coleman died on the M25 that foul night, Abigail Gould received her lungs, making her the oldest recipient of one of eight organs that Sophie donated.

4

Jack Gibson

While Abigail Watson's family were working class, Jack Gibson's family were relatively poverty-stricken.

He was born into a family that lived in a poor area of London's east end just after England won the World Cup in 1966.

There were six members of the Gibson family, who all lived in run-down council accommodation with only three bedrooms. Jack Gibson had one brother, with whom he shared a bedroom and two sisters, who also shared a bedroom. Jack was the second eldest of the four children.

Reg Gibson, his father, worked as a labourer on the docks along the Thames. It was hard, menial work for little money. His mother had been a tea lady before she married and had children. She stayed at home looking after the children. The father's income kept them just about afloat. Dot (Dorothy) Gibson took in occasional laundry for some extra income that was never enough to give them anything but a spartan lifestyle.

Two adults and four children living in a run-down

council house with three bedrooms and a small income did not make for easy living. The siblings got on well enough, but that did not mean that they did not argue or fight.

Jack's childhood was blighted by poverty and parents who were strict to the point of being brutal. All four children were familiar with a clip around the ear and not just from their parents. It was not unusual for the police, shopkeepers and traders in that part of London to use physical violence routinely on misbehaving children. No use crying to your parents as this was what they experienced in their childhood. If it was good enough for them, it was good enough for their children.

Jack's schooling was unremarkable except for his attendance, which was patchy at best. He would bunk off school regularly with some friends. He became part of a gang of school-age children who committed petty crimes in their neighbourhood. The most likely long-term outcome for Jack was that he would become a career criminal who might spend much of his time dodging the police or serving time in prison.

In 1985, Jack joined the Army. It was not a choice he would willingly have made had he not needed to get away from the Metropolitan Police. He was out of work, and the police suspected him of being involved in a violent robbery of an Asian shopkeeper. Despite their best efforts, they could not find enough evidence to link him to the crime. Jack was interviewed about the crime but was released without being charged. He assumed it would only be a matter of time before the Metropolitan Police 'felt his collar', as he was heard to say.

Jack had a chance encounter with a soldier at his local pub shortly after his police interview. Their talk convinced Jack that an army career was his way out of his chaotic lifestyle.

The squaddie told Jack he got three meals a day, bed and board. The pay was not great, but it bought a few beers on a night out. They also supplied him with his uniform and every item of clothing he would need.

This was a much better prospect than he could have if he stayed around the east end of London. The soldier told him he was on his first leave after basic training that was all done in Catterick, in Yorkshire.

Three meals a day plus bed and board, with enough money for a few beers, seemed a splendid idea to Jack. Better still, Yorkshire was hundreds of miles from the Met Police. Maybe they would not bother to chase him up there. He had also seen the hero's welcome given to the soldiers returning from the Falklands War. He fancied himself as a hero. The number of dead soldiers who would never return from war did not register with Jack.

The following day, he visited the nearest recruiting office and enlisted in the Army. Due to his limited education, he had few options for army positions and becoming an infantry soldier, also known as "cannon fodder" by some.

This decision was typical of Jack. Taken without much consideration. If it seemed like a good idea at the time, it was his usual modus operandi.

At Catterick, the Army taught their recruits many skills to make them good soldiers. Among other things, Jack was taught aggressive fighting skills and respect for their superiors.

The wilful child and rebellious teenager that was Jack Gibson always had difficulty with authority figures. His issue with fighting was that the Army needed him to display it at their pleasure, not his. Fighting in pubs or the NAAFI bar wasn't a good reflection on the Army. Jack was well-versed in all of his Commanding Officer's defaulter parades. There were fines and occasional spells in military prisons throughout his service. He also spent several nights in local civilian police cells after rowdy nights in nearby towns.

After his twenty-six weeks of initial training, Jack was posted to the Royal Regiment of Fusiliers. He served two tours in Northern Ireland in 1987 and 1993 and Kuwait during the 1991 Gulf War.

The fighting in Iraq and Kuwait was punishing because of the heat and sand. At least the enemy was easy to recognise.

It was a different story in Ulster. The enemy looked just like him and every other British citizen. While there was less actual fighting, the strain of not knowing who the enemy was became a constant source of anxiety. During each tour in Ulster, Jack and his comrades were in continual fear of assault from every quarter.

The British Army was there to perpetuate the fiasco of the 1922 partition of Ireland. This led to a Catholic

Irish Free State and a segregated protestant Ulster, each hating the other.

There were Catholics living in Ulster, but they were marginalised by the Protestant government. There was extreme social inequality in every aspect of Catholic life in the province.

Peaceful Catholic civil rights marches in 1968 ended in violence, and the Provisional IRA saw this as an ideal opportunity to use to their advantage. The protestant community responded with their versions of citizen armies, so the 'troubles' began.

Jack had no interest in the politics of the Irish troubles. The Army was there to keep peace between warring Catholic and Protestant factions. The problems were exacerbated by the fact that each one hated the other two in equal measure.

Jack left the Army in 1997, long before his regiment was again sent to invade Iraq in 2003. His discharge date was delayed by nearly eight weeks. In the armed forces, time spent in military prison was added to the end of their service contract. Jack got his campaign medals, but if he had served long enough, he would never have received a Long Service and Good Conduct Medal. If the clerks had seen Jack's disciplinary record, they would have most likely laughed.

Jack would have liked to remain in the Army after his twelve-year contract ended. Unfortunately for him, his many brushes with military and civilian police made that impossible. The Army wanted men who fought on the battlefield, not in their barracks, NAAFI or local pubs.

Jack found it difficult to cope with life outside the Army. The Army had fed, clothed and housed him and given him enough money to buy a few beers at the end of the week. Life in civvy street meant he had to fend for himself. Jack discovered he was not good at that.

Jack met Amy Ramsden in 1998. Amy and Jack were well matched in that both had strong views and were unafraid to voice them. Both had menial jobs. Jack was a jobbing labourer, and Amy worked behind the counter in his favourite greasy spoon cafe.

It was no surprise to anyone that the two married in 1999, just before the start of the new millennium. To both, it seemed like a good idea at the time.

Two daughters followed swiftly, but all was not well within the Gibson family.

Veterans from the armed services knew about PTSD, but in keeping with their aggressive male attitude, most thought others might have it, but not them. Jack did have PTSD, but it was undiagnosed because he never bothered to seek help.

Jack started to drink and found it increasingly difficult to keep a job as a result. His wife was initially tolerant, but his drinking became worse. He became more aggressive and started to be violent towards her and his two daughters. As a result, his family was permanently in debt and afraid. Eventually, Amy decided that she and the girls had taken enough, and she left him to live with her parents in Shropshire. Jack stayed in their home in Lambeth.

When Amy sued for divorce, Jack did not contest it, nor did he take any responsibility for the breakup of his marriage. Discipline in the Army was strict. His wife and children needed discipline. The odd slap around his ears when he was a boy did him no harm, reasoned Jack. He was only acting as he saw fit. He was the husband and father, and he was responsible for their discipline. What had been good enough for him was good enough for them. Amy disagreed, and as a result, she sued for divorce as soon as she could.

The separation only made Jack's situation more difficult and his drinking worse. He was in and out of work. He never had enough money to pay what was due on his child support or his rent.

His continued drinking made him tired and weak. He also lost his appetite, and he developed jaundice. When his skin and eyes turned a pale yellow tint, he consulted his GP.

"I look like a Chink," complained Jack in his usual racist language.

It did not take the GP more than a few minutes to conclude that Jack's drinking was the root of his problem. He could smell it on his breath as he spoke. He bluntly told Jack that he needed to drastically change his lifestyle. He told Jack to stop drinking and to maintain a healthy lifestyle. He could start his healthy lifestyle by avoiding fatty foods. The GP clearly had little sympathy for a man who might want to drink himself to death.

The question of Jack's time in the Army and possible PTSD was never mentioned, and, as a result, his habits did not significantly change. His liver deteriorated, and the only treatment his GP could prescribe had been ignored.

Jack was eventually referred to his local hospital, where he met Mr Keith Isaac, a consultant at the King's College Hospital in London. As a result of his first consultation, Mr Isaac referred Jack for treatment of his PTSD in the hope that this might help him stop drinking. Jack found it challenging to ask for help or discuss his problems. The therapy had limited success.

Eventually, Mr Isaac decided that Jack's condition had deteriorated to the point where he needed a transplant.

The option was to stop drinking and be put on the transplant register or keep drinking and die soon. Mr Isaac told Jack point-blank that he would not be placed on the transplant list unless he gave up drinking. He explained that he had no intention of spending a vast amount of tax-payers money, only for it to be wasted by a drunk.

"Give up the booze and I'll put you on the list," said Mr Isaac at one appointment. "If you do not give it up you will slowly deteriorate and die. The choice is yours."

Jack decided that he was not yet ready to die and told his consultant that he would stop drinking and join Alcoholics Anonymous.

"We'll keep an eye on you Mr Gibson. When I'm convinced that you have heeded my advice, then we'll talk about a transplant. If you keep drinking, I'll manage your decline until you die." This was what Keith Isaac called a sobering ultimatum.

This starkness of the warning shook Jack to the core. He did attend AA meetings, and he did satisfy Mr Isaac that the money the NHS would spend on him, should a liver become available, would not be wasted.

While his doctors were still considering if he should be placed on the transplant register, Jack asked an awkward question.

"If I get a transplant, will you tell me everything about the donor before I agree to take it, or not?"

"You have all the information that you need about a possible donor," said Keith Isaac.

"You've told me nothing."

"Exactly. I only get information about the health of the organ and any possible issues in the donor's health that might be relevant. That's all that's needed."

"What if I insist on knowing stuff about the donor?'

"You can insist all you like the answer is the same."

"I might still insist."

"In that case I will annotate your notes to that effect and you will never go on the transplant list. I will monitor your deteriorating condition until you eventually die. The choice is yours."

Jack withdrew his question and agreed to his consultant's conditions and advice. He continued with his regular AA meetings.

"Should we put this man on the transplant register?" Asked Sanjay Pather, Keith's registrar.

"He has done all that I asked of him and we have followed the protocols. The next logical step is a transplant," said Keith.

"Surely there are other more deserving candidates for a liver."

"Perhaps, but that is not our choice to make."

"The man's a racist."

"That he is but refusing him an organ on that basis means that we are practicing eugenics, just like the NAZI's did during the war. Everyone deserves a second chance."

"Even a bigot and racist?"

"Even bigots and racists. He may not get one in time to save his life. As you know, demand outstrips supply of organs. Whether he gets one or not is in the lap of the gods. Did I just compare the transplant co-ordinators in Bristol to gods? Does that constitute blasphemy or heresy?"

"Both, but I'll pray for you during my next visit to my temple," said a smiling Sanjay.

"Thanks."

Jack was eventually placed on the liver transplant list in 2012. He would get Sophie's liver.

5

Rose Prentice

Rose Prentice was born Rose Ramsden in Chelmsford, Essex, in 1967. She was the outcome of her father and mother's celebration of England winning the World Cup. Her father, Bert, had hoped for a boy so that he could call him Bobby, after Bobby Moore, the England captain of the winning team. Unlike Bobby Moore, Rose's parents were disappointed. Rose's father still thought that Bobby could double as a girl's name, but her mother, Margaret, was having none of it.

"You'll be wanting to call her Alfredia after the manager next," said Margaret. "If you want her to have a name associated with England football, why not call her Rose?" And so, it came to pass that the girl was named Rose.

Rose's parents were middle class and comfortably off. They gave her and her younger sister many toys other children longed to have. The girls were encouraged by both parents to work hard at school. This meant that they would have more career opportunities upon leaving.

Rose's teenage years were a time before mobile

telephones. The internet superhighway did not exist either. The most sophisticated forms of information searches of this time were done on Ceefax or Teletext. To make friends actually meant talking to people face to face. Social media did not exist. Rose did have a pen pal in Australia. Rose partnered with her pen pal through a teen magazine that encouraged it.

She needed to create an advertisement seeking someone with similar interests. She got a reply from Jenny Matthews in Sidney. The pair wrote letters to each other for about eighteen months before the whole thing petered out.

The letters from Australia were full of the beauty and sights of the country. Rose was envious as she longed to travel to exotic places. She saw photographs and read travel stories in magazines and newspapers that fired her imagination. Rose would be a world traveller in her dreams.

Rose attended Broomfield Primary School before continuing her secondary education at Chelmsford County High School. Although she was not the top student in her class, neither was she a poor performer. Rose gained several above-average O-Level grades but chose not to pursue A-Levels. It was a decision that she would later regret.

Despite being English, Rose loved the music of the Scottish band called The Bay City Rollers. Rose was a member of the Tartan Horde, the name given to English fans of the band. She also liked the music of Rod Stewart, who was born in London to parents of

English and Scottish descent. Her father was horrified as Stewart was an ardent supporter of the Scottish football team. He viewed Scotland as the Auld Enemy. Rose had no interest in football. Her only passion was the music she listened to, and it just happened to be produced by Scottish artists.

Rose also became addicted to the Australian soap opera Neighbours. She and Jenny both wrote about their love of the soap opera. Rose got advance notice of the plot, which arrived with Jenny's letters before the show aired on UK TV.

Like many young girls of the time, she eulogised about the wedding of Prince Charles to Lady Diana Spencer. Rose was a great admirer of the bride but was heard to wonder why the beautiful princess would marry a frog? Rose was heard to say that if the Lady kissed the frog more often, he might become a handsome prince. Nevertheless, she collected the commemorative coins celebrating the wedding.

When she left school, Rose got a job as a junior clerk in a solicitor's office in the town. She attended a local secretarial college in the evenings to study shorthand, bookkeeping and typing. This was still a period when a woman's role was to be a wife, mother and homemaker. Having a good brain was not necessary for that role in life.

While there, she worked with a junior lawyer named Derek Prentice. The two started dating, and they eventually married in 1989.

After their marriage, Derek set up his own practice in Wood Green, London. The couple was thrilled to start their new life together, with a new house, business, and plans for a family. They had a son in 1992 and a daughter in 1994.

Shortly after their move, Rose received news of her mother's serious medical condition. She had been diagnosed with Autosomal Dominant Polycystic Kidney Disease for which there was no cure. The joy of her marriage and her husband's new business was tempered by the fact that she was not as close to her mother as she now needed or wanted to be. Her sister was even further away in Manchester.

Rose's mother had been slow to consult her doctor. Her diagnosis came as a complete shock to the whole family. Her mother had the classic symptoms of the disease, such as abdominal pain, multiple urinary tract infections (URTI), and high blood pressure. Her GP had previously checked her blood pressure. It was elevated but not high, in his opinion, and so he prescribed propranolol. No further action was taken. She had several prescriptions for antibiotics for her URTI, but again, no further action was taken.

When she detected blood in her urine once again, she contacted her doctor. After his initial examination and seeing the results of the tests he ordered, he thought it was time to refer Margaret to the local hospital.

The diagnosis was made too late, so Margaret's treatment focused mainly on relieving her symptoms. She was ordered to stop taking ibuprofen for her pain as it was known to disrupt kidney function. Margaret had the added severe pain associated with occasional kidney stones. During these painful episodes, Margaret thought she was dying.

As her disease progressed, she was placed on CAPD but died in 2004 before she could receive a kidney transplant, which had become her only option. Margaret was just 60 years old, and Rose was thirty-seven.

Rose was devastated, not just by her mother's death but also by the knowledge that this disease could be inherited. She only learned this after her mother's death. She consulted her GP because she felt tired, had lost weight, had muscle cramps and swollen feet.

The doctor examined Rose and obtained her mother's medical history from her. He ordered a standard blood test. Upon reviewing the results, he referred Rose to the nearby renal unit. The blood test revealed that Rose had inherited the same kidney disease as her mother. This was the point that Rose became aware of the familial connection.

Rose became a patient of Mrs Isabel Mason at The Royal Free Hospital in North West London.

After the physical examination on the initial appointment and history taking, Mrs Mason ran a series of tests on Rose, which included an ultrasound and CT scans. These confirmed the diagnosis of Autosomal Dominant Polycystic Kidney Disease. She asked Rose to come into the hospital as a day case patient for a kidney biopsy. Rose agreed.

Mrs Mason also learned that Rose took over-the-counter ibuprofen for her constant back pain. She was told to stop that immediately. The consultant explained the reasons for the instruction. She was told they could disrupt kidney function and interfere with medications that would be prescribed to control her blood pressure.

Rose had no idea that a simple over-the-counter medicine could have such a devastating effect on her kidneys. Why would she? Over-the-counter medicines were safe. That's why you could buy them without having a doctor's prescription. Her mother never mentioned this problem during their conversations.

The consultation following the biopsy was difficult for the patient and the doctor. Mrs Mason did not like to tell her patients that they had an incurable disease that had been inherited from their mother. Rose had the experience of her mother's decline and death and saw the same path for herself.

The diagnosis was a double blow for Rose. She had a severe medical condition that prophesied a premature death. She and her husband were beginning to get their finances in order. They had discussed taking their first foreign holiday as a family. Foreign travel was put on hold indefinitely. Medical insurance premiums might cost her more than the rest of the holiday. Derek's practice was doing quite well, but he was still trying to expand his client base.

Just like her mother before her, Rose experienced occasional kidney stones and the excruciating pain they caused. She had two treatments of shock wave therapy to break the kidney stones down to allow an easier path to flush them away. She was also not a stranger to antibiotics that were given for her bouts of urinary tract infections. As if period pains were not enough, she was heard to complain.

Rose's blood results continued to decline to the point where she was placed on Continuous Ambulatory Peritoneal Dialysis (CAPD) three times a day.

Despite following the treatment regimen of Mrs Mason, Rose's condition deteriorated, as she knew it would. When she reached stage five of the disease in 2012, Rose was put on the transplant register for a kidney.

On 12 February 2013, Rose received one of Sophie's kidneys.

6

Megan Wilson

Megan Thomas was born in Nottingham in 1970. Her family were originally from Wales but had moved to Nottingham before the birth. The family lived in the Hyson Green area of the city.

Megan's father, Gareth, had a pronounced Welsh accent like his wife, Brenda. Born in England, Megan did not. She almost felt like she was in a foreign country when she was at home. If her parents did not want her to know what was being said between them, they spoke Welsh. Megan did not want to learn how to speak Welsh. Where would she use it except at home? Trips to Wales to see members of her parent's families were few and far between. There was no incentive to learn as they stayed in homes with older people with no one of Megan's generation present.

The 1980s was a decade where the yuppies were prevalent. They were ambitious and often driven by greed. Yuppies personified the free market ideals that prevailed during the political ascendancy of Margaret Thatcher and the New Right. Greed was good: greed was right.

At the other end of the scale were black British youngsters who were particularly hard hit by unemployment and inner-city deprivation. The hardships they faced detonated some of the worst urban violence in British history. There were riots in Liverpool, London, Cardiff, and Oxford.

The 80s were an age of technological development with the Apple and Sinclair home computers, plus the first mobile telephones nicknamed the brick because of their size. They needed a shoulder bag to carry them.

Megan's first school was Claremont Primary, where she did reasonably well before moving to the Nottingham Girls Academy for her secondary education. She left school with two As and five O-level passes at decent grades. When Megan was young, she would cover her school books with sticky-backed plastic that she purchased from Woolworths, just like many other children.

During her teenage years, Rose was a big fan of the pop group Duran Duran and the music of David Bowie. At age twelve, Megan became addicted to a TV show called Fame. It was a story about young students at the New York City High School of Performing Arts. The show was a massive hit in the UK, so Megan was not the only teenager hooked on the series.

After finishing school, Megan started her job hunt. She decided to apply for a position at Boots, the Chemists, as it was one of the biggest employers in the city. She was given the role of an apprentice and received training from the company in merchandising. Her apprenticeship was based at the company headquarters located in Beeston.

In 1992, Megan met Eric Wilson at the Beeston site where they both worked. Eric was a pharmacist who worked in the research department of the company. However, their relationship started badly when Eric accidentally spilt some hot drink on Megan in one of the canteens. Despite his mistake, Eric was very apologetic. After some hesitation, he offered to take Megan out for a nice meal. He also offered to pay for her dry cleaning. Megan accepted and found that she got on well with Eric.

The relationship progressed. Eric was a kind, loving man who was always attentive towards Megan. She often thanked the spilt hot drink for introducing him to her. The couple married in 1994.

They rented a house in Nottingham while they looked for somewhere to buy. Just as they were about to settle on a house purchase, Eric got the offer of a place to manage the pharmacy at their big store in Oxford. It was a promotion, and it came with a significant pay rise. It would make finding a good house that they could afford easier.

The couple decided to buy a new house in Oxford instead of Nottingham. Megan acknowledged that her skills in merchandising were valuable and that she could still work for Boots or any other big high-street outlet.

Having settled into their new jobs and home, Eric booked a holiday at the Grand Hotel in Sharm el-

Sheikh, Egypt, for their third wedding anniversary. Unfortunately, Megan got a dose of The Pharaoh's revenge. Despite Eric's qualifications as a pharmacist, Megan needed the expertise of a local doctor to resolve the issue. The problem was resolved within two days, and the rest of the holiday was enjoyable.

Sometime after she got back to Oxford, Megan had another gastric upset. Eric's expertise with medicines helped alleviate the symptoms, but they recurred several weeks later. Two more gastric upsets in quick succession prompted Megan to seek advice from her GP.

The general practitioner prescribed a more powerful treatment than Eric was qualified to provide. Although it worked for a while, the upsets persisted, and the GP remedies became even more powerful. Eventually, the GP referred Megan to the Churchill Hospital for further investigations. It was at the Churchill Hospital that Megan met Mr. Kasara Doshi.

Mr Doshi's examination raised strong suspicions that Megan had Crohn's disease. Her foreign travel and gastric symptoms reinforced his view, and he ordered a series of tests. These tests required Megan to give blood and faeces samples. She also had an ultrasound scan and x-rays.

When he received the results of the tests, Mr Doshi then asked for a CT scan. This confirmed his diagnosis and gave him a clearer picture of where the disease was situated. He initially prescribed steroid tablets for Megan. This only had a minor effect, so he moved her onto a course of immunosuppressants.

Over several years, Megan continued to have routine

hospital appointments in Mr Doshi's clinic, but her condition continued its slow decline.

Mr Doshi ordered an endoscopy on Megan's gut and suggested a biopsy. She reluctantly agreed and spent a day in hospital for the procedure.

Not content with pushing a tube down Megan's throat into her gut, Mr Doshi gave her the added ignominy of a tube going the other way when he performed a rectal colonoscopy examination.

After undergoing all of these tests and taking various medications, Megan's disease was not being managed successfully. Her condition was deteriorating. Despite the efforts of everyone, Megan's symptoms were getting worse rather than improving. This made her feel almost relieved when Mr Doshi suggested a bowel transplant as a possible solution to her problem.

The final decision would be made after another series of tests. Some of these tests were done as an outpatient, but she also had to spend five days in hospital for others.

In the hospital, Megan and Eric met the transplant team. The team were contemplating a small bowel transplant, which was a long and complex operation, lasting anything up to ten or eleven hours. This was the option she was given and was recommended because she did not have associated liver disease. The complications of infection were emphasised as the intestine is not a sterile organ.

After a prolonged discussion with Eric, Megan agreed to proceed if an organ became available. Meanwhile, she would continue on her current treatment regime. Megan was listed for an organ transplant in 2012.

On 12 February 2013, Megan received Sophie's small bowel, all sixteen feet of it.

7

Arthur Morgan

Born in 1979 in Cardiff, Wales. Arthur's father, Bryn, was a Cardiff City Council Development Manager working in education.

Arthur was still young when it was discovered that he had poor eyesight. He was having trouble at school seeing the blackboard and his books. His teachers thought he was slow at first. The school soon worked out it was his eyesight, not his brain, that was the problem. It was not dramatic but required spectacles to enable him to read and do his schoolwork.

Arthur began schooling at the Cathedral School in the city as a junior and then as a senior pupil. He managed six GCSEs before moving to the sixth form to study Computer Science, Maths and Physics as his A-level subjects. Arthur went to Bath University to study criminology, gaining a 2:1 degree.

At school and university, Arthur had played rugby. He played the sport despite wearing glasses he only needed for small, close work such as reading. Like most Welshmen who played rugby, Arthur dreamed of running out onto the hallowed turf of Cardiff Arms Park, now the Millennium Stadium. As an average player with poor eyesight and several allergic conditions, he knew it was only a dream. Still, despite his poor eyesight, Arthur could see himself in the red jersey running onto the pitch and singing Hen Wlad Fy Nhadau or Land of My Fathers for non-Welsh speakers.

The problem for Welsh rugby in the 1990s was that the Welsh team of that era were a shadow of the glory team of the 1970s. Arthur often found himself thinking that he could play better than some of the players on the Welsh rugby team, even with his glasses on, as he watched them suffer another humiliating defeat. Unfortunately, Welsh rugby suffered from a lack of funds, and several of their best players had left for lucrative contracts with rugby league teams in the north.

Arthur used his degree from Bath to get a job as a social worker in London. His post was based in central London. Arthur rented a flat in Barnet in North West London. He had a short walk to Hendon Central Tube station to take a train direct to Euston. From there, it was another short walk to his office.

Working in inner city London, Arthur saw much of the worst of human behaviour. It always brought a smile to Arthur when he witnessed the generosity and humanity that was also equally prevalent in the area. He often struggled to understand why some people were so awful while others were full of compassion and kindness. It was difficult for him when he saw both in the same household.

Arthur was six feet two inches tall, thickset and broad-shouldered. He looked like he should be a rugby player but had given that up when he moved to London. He had gone to join a club but found the culture in the club elitist. Rugby was mainly a sport played by working men in Wales. In the suburbs of London, it was a gentleman's sport full of toffy-nosed snots who spoke with plumbs in their mouths. A bunch of 'Hooray Henry's' was how Arthur described the club membership to his family. Arthur felt out of place in their company, and so he left.

He kept up his fitness by joining a local gymnasium. While there, he met Amanda Gregory, one year his junior. He asked her out, and the relationship blossomed. The pair were married in 2003.

The couple bought a house in Hendon when they decided to start a family, as they realized that a small flat was not a suitable place to raise children. They had a daughter in 2006 and a son in 2008.

Arthur's vision continued its slow decline as he aged. The prescription in his spectacles became higher after each regular eye test. As his vision deteriorated, he required regular changes in his lenses. It was a necessary, expensive need. During one of his routine tests, his optician suggested sending him to Moorfields Eye Hospital for examination. She was concerned that his vision was getting worse, and she obviously suspected his underlying condition but did not elucidate to Arthur.

At the end of his initial Moorfield examination, the consultation told Arthur he was suffering from a condition known as Keratoconus in his left eye. He might as well have been talking Swahili for all Arthur knew. Arthur asked for an explanation.

"What is this Keratoconus? How did I get it?"

"Keratoconus is a non-inflammatory eye condition in which the normally round dome-shaped clear window of the eye, the cornea, progressively thins causing a cone-like bulge to develop. This eventually impairs the ability of the eye to focus properly."

"As to the question of how you come to have this condition, the short answer Mr Morgan, is that we do not know. There is a theory that there may be a genetic link to the condition, but we are not sure. What we do know is that it is more common in people with multiple allergic conditions, such as eczema and asthma, all of which you have."

"What can you do for it?"

"There are two possible treatments. One is Corneal cross-linking and the other could be implanting intrastromal corneal ring segments."

"Will these two treatments cure me?"

"No, they only slow down, or possibly stop the condition getting worse."

"So, what's your suggestion?"

"Your condition is more advanced than I might have expected at this first consultation. I'm surprised your optician did not refer you sooner. I'd like to monitor it for a little time and decide what to do if it continues to get worse."

"What is your assessment at this time?"

"I think we may have to consider Corneal cross-linking as the best option, but let's wait and see."

Arthur knew about Crossrail and the circle line but nothing about Corneal collagen cross-linking or intrastromal corneal rings. His chemistry and physics studies gave him no insight into optical conditions. He asked the consultant to explain what he meant.

"The procedure is done as day case surgery under local anaesthetic. After the anaesthetic drops take effect, you lie back and small clip is placed in your eye to keep it open. We then gently brush away the epithelium of the eye before they are saturated with riboflavin eye drops. These drops are applied every few minutes for at least ten minutes. Then, we shine an ultraviolet light into your eye for eight minutes. When we are finished, we attach a soft contact lens to your eye. This is a bit like bandaging a wound."

"Then what?" Asked Arthur.

"Well, you keep the lens in for about a week until the next follow up appointment. If it falls out, don't try to replace it. You might do more harm than good."

"Will I need to take time off work?"

"Yes, you'll need a week at least."

After two more follow-up appointments, Arthur's consultant suggested it was time to consider Corneal cross-linking. His condition was getting worse.

"The condition in your left eye is slowly getting worse, Mr Morgan. I think it's now time to move to the next stage. The right eye seems to be stable, but we'll keep an eye on it, if you'll excuse the joke."

"Do what you have to do, Doctor," said a resigned Arthur.

Everything was arranged, and the procedure went ahead as planned. Arthur took a week off work as agreed. The contact lens did not fall out before his next appointment when it was removed.

Unfortunately, the treatment did not halt the progress of the condition in his left eye, and Arthur was advised that a Corneal Transplant was now the only viable option for that eye. The other eye had signs of deterioration but was stable, at least for now.

Arthur was placed on the cornea transplant register in late 2012.

On 12 February 2013, Arthur received Sophie's right cornea.

8

Jacob Marshall

Jacob Marshall was born in Watford in 1989 to a middle-class family. His childhood in the London suburbs was filled with news of IRA bombings and even more new skyscraper buildings being built in the city.

When he was ten years old, Jacob was alive to see the opening of the Millennium Dome. It was situated on the south bank of the River Thames. This was a major project designed to mark the arrival of the third millennium. However, after the exhibition ended, the government did not know what to do with the expensive structure. It took another five years before it was rebranded as the O2 Arena. Jacob occasionally attended music events at the venue, still widely known as the Millennium Dome.

As for culture, the 90s were a time of cyberpunks, yuppies and mobile telephones the size of breeze blocks.
Perhaps the biggest event of the decade, or century in some people's minds, was the death, in 1997, of Diana, Princess of Wales. The country went into an outburst of

mourning, the like of which had never been seen before or since. Jacob was not one of the mourners. He was no lover of the Royal Family.

Jacob started primary schooling at Nascot Wood Junior School before moving to Watford Grammar School for boys.

He was a passionate follower of his local football team, Watford FC, locally known as the Hornets. Jacob regularly visited their small Vicarage Road stadium to watch home games. Jacob and some friends would travel to away games whenever they had the money to pay for the trip. Unfortunately, his team was not doing well for the most part. The club's glory days were five years before Jacob's birth, and since their famous manager, Graham Taylor, left, their fortunes had been in decline. However, Taylor's return in 1997 as the manager brought about a change of fortunes, and the team started rising again. Jacob's loyalty to his hometown club was like a rollercoaster ride as he experienced the highs and lows of the team's erratic performance.

Jacob was pleased when he left school in 2006 with good exam results. He was even more joyful when his football team won the playoff final against Leeds United, promoting them back to the Premier League. Double joy for Jacob.

Jacob got a job with a large accountancy firm in the town. He went to further education classes to study accountancy.

Jacob's politics were slightly left of centre. He despised the greed that engulfed the city of London and much of the Western world. He was not keen on the Royal family and had toyed with the concept of Britain becoming a Republic. Friends joked that he might one day lead the revolution to depose the Queen and replace her with what? That was a problem because he could not see Tony Blair, or any politician, as President of the country. It was almost as if the Royals were the lesser of two evils.

Jacob started to have symptoms of tiredness and occasional shortness of breath in his last two years at Watford Grammar. He put these down to the pressure of work leading up to his A-level exams. Jacob thought his symptoms were related to the stress of his examination preparation rather than a medical condition. He consulted no one about his problem.

A year or two after he started work with the accountancy firm, he began to have chest pains as well as his other two symptoms. He was uncertain whether the pains he was experiencing were associated with his preparation for the accountancy exams. He was still living with his parents, who persuaded him to seek the advice of his GP. He made the appointment.

The GP started his examination by listening to Jacob's heart and lungs with a stethoscope. He took Jacob's blood pressure.

After Jacob responded to his initial questions, the GP dug deeper into his family history. Had any of his family had these symptoms?

"Not as far as I know," replied Jacob.

"What about blackouts or palpitations?"

"I've not had blackouts but I have had palpitations

during my A Levels and accountancy studies. I put it down to stress."

"I think we need to do an ECG to check your heart," said the GP. The mention of a test to check his heart shook Jacob. He was not expecting that. "Can you make an appointment with the practice nurse for the test?"

"What does and ECG do?"

"It's a simple test to check the rhythm and rate of your heart."

"Do you think there's something wrong with it?"

"It didn't sound quite right when I listened. I'll need this test to see if there might be an abnormality."

"OH."

"Make another appointment to see me after the ECG test and we'll discuss what we have found, if anything."

The appointment following the ECG raised Jacob's level of anxiety.

"Your blood pressure is high and ECG suggests that further examinations are needed. I'd like to send you to the hospital for a cardiologist opinion."

"A cardiologist, does that mean there's something wrong with my heart?"

"There may be, but they are the experts and will be able to give you a clearer picture of any issue. I'll send a letter today."

Jacob left the surgery feeling anxious. He had something wrong with his heart that needed the expertise of a cardiologist. This did not sound good. In his anxiety and confusion, he had also failed to question his GP for further information.

While he waited for his initial hospital appointment at Watford General Hospital, he searched the internet.

This did nothing to relieve his anxiety as he learned that cardiac problems often resulted in premature death. The thought of dying young had never previously crossed Jacob's mind.

The cardiology consultant looked at the GP's ECG results and Jacob's blood pressure. He decided to run an exercise ECG and MRI. These tests showed that Jacob did have a cardiac disorder that needed treatment and to be managed. He indicated that Jacob had dilated cardiomyopathy.

"What this means, Jacob, is that is a disease of the heart muscle which makes the muscle walls stretched and thin. That's the dilated part of the disease. The thinner walls are weakened, and this means the heart cannot contract properly to pump blood to the rest of your body," said the consultant.

"How did this happen?"

"In your case we just don't know. The possibilities are an uncontrolled high blood pressure. Your pressure was high, but it is being controlled fairly well by the medication your GP prescribed."

"If not that, then what?"

"You might have a disease of your blood vessels or you might have had a serious viral infection in the past. I see no record of anything in your GP note to me."

"I did have something as a kid, but I was told it was just one of those usual childhood illnesses."

"It could be as simple as a lack of vitamins. What's your diet like?"

"I try to eat healthy foods and stay away from fast food as often as I can."

"What about alcohol or recreational drugs?"

"I don't use drugs, unless they are prescribed by a doctor. I probably drink quite a lot but I don't often drink to excess."

"In that case, Jacob, I just don't know."

"What's the next step?"

"I think that an angiogram is needed." The doctor went on to describe what happened during that procedure. It sounded unpleasant to Jacob, but it seemed necessary. He was prepared to try anything to avoid dying young.

Jacob returned for the angiogram, which was not as unpleasant as expected. The outcome was a change in medication and a suggestion that further treatment might involve implanting a device into his chest. Jacob was informed that he needed to tell the Vehicle Licensing Authority about his medical condition since it might restrict his ability to hold a driving license.

This news came as a great shock to him. He was also advised to avoid engaging in extreme physical exercise. Fortunately, his sedentary job did not require much physical exertion. However, when he was at football matches, he would jump up and down while waving his arms, cheering or booing.

Unfortunately for Jacob, his condition deteriorated, and he had a pacemaker and a cardioverter defibrillator implanted at Harefield Hospital. The condition continued to deteriorate, and he was put on the heart transplant list in 2012.

On 12 February 2013, Jacob received Sophie's heart.

9

Richard Ainsworth

Richard was born in Sutton, Surrey in 1993. He was the eldest of three children born to Joyce and Edward Ainsworth.

From an early age, Richard was a sickly child. The first indication of his medical difficulties appeared when he was six months old. Over a long weekend, his temperature fluctuated dangerously, spiking and dropping unpredictably.

There were no clear indications of what could be causing the fluctuations in temperature. The family general practitioner relied on the standard treatment for suspected inflammation, which involved administering antibiotics in escalating doses. However, after four days, the GP called for an ambulance, and Richard was immediately taken to the paediatric unit of the local hospital.

After undergoing multiple examinations and tests, it became evident that Richard's condition was quite severe. The consultant diagnosed glomerulonephritis, a disease of the kidney.

The consultant could not identify a cause, and it was

called idiopathic. This is the medical-sounding term for we do not know. Sometimes doctors even use the term GOK (God Only Knows).

Richard's condition was diagnosed as being progressive and without a known cure. This came about after the hospital took a biopsy to assess the condition. After examining the biopsy, the doctors could only offer help to slow the progress of the disease. Unfortunately, the long-term prognosis was not good. However, he was not required to stay in the hospital and was discharged.

The hospital doctor advised Richard's parents to help him lead as much of a settled life as possible. His parents were given guidance about his dietary and fluid intake. They were warned it would not be easy.

The difficulty of feeding a small child with the foods recommended by the hospital soon became clear to his parents. Richard did not like the taste of gluten-free food. Joyce and Edward tried to motivate him by eating it themselves. They wanted to show how much they enjoyed it. It was a false enjoyment because the stuff tasted like cardboard to them. Still, Richard got used to it, and the quality and taste of what was on offer improved over time.

He was also advised to reduce salt, fat and potassium in his food. Exercise was good for his condition. Richard avoided becoming overweight by following medical advice and engaging in regular sports activities.

Richard attended school like all children, but his condition continued to steadily deteriorate. He started his primary education at Westbourne Primary Academy before moving to Sutton High Senior School.

Throughout his school years, he was able to participate in physical education activities. Still, he had to frequently visit his local hospital for routine check-ups due to his medical condition. Over several years, his blood and urine test results gradually declined. As a result of his condition and frequent hospital visits, his academic performance was also affected, and he was only able to achieve a few GCSEs with average or slightly above-average grades. He did not study for A-level exams. Before he made that decision, he wondered about going to university. How could he cope with his deteriorating medical condition while studying for a degree? Thinking about the disruption his condition might bring and the cost helped him decide not to proceed.

He had left school and was now at work in the office of a local data storage company.

As his condition deteriorated, Richard was referred to the specialist unit at St George's Hospital. He came under the care of Mr Allan Cunningham.

In his early twenties, his test numbers had deteriorated to the point where a more radical medical approach was required. It was then that his consultant indicated he would need Continuous Ambulatory Peritoneal Dialysis (CAPD).

Richard and his family were made aware of the medical pathway before him. It was bleak. They were regularly updated on how close he was getting to the next stage of his treatment.

At an early age, his family had been told that the first stage of his treatment would be a controlled diet and continuous monitoring of his blood toxins. Stage two would probably be CAPD. Stage three would be haemodialysis. The final option was a transplant or end-stage kidney failure leading to an early death.

"Your numbers are high and we are unable to keep them down with the current regime. It's time to move to CAPD," said Mr Cunningham.

"Is there no other option?" Asked Joyce.

"I'm afraid not. In our recent consultations, we have talked about the numbers and how they are getting worse. You, we, are now at a point where we have limited options. CAPD is better for you than haemodialysis. It means you can still work. As you both know, haemodialysis has to be done in hospital and takes up most of your day. You can have CAPD treatment when you get up and again at night when you get home. I'm sure your employer will help you to have a clean safe place for the third one during your lunch break."

"They have said so. What about a transplant?" Asked Richard.

"Your numbers are not high enough for that. There is a large waiting list and there are many more people with a greater need than you at this time. You do not yet qualify to be put on the list, even at the bottom. As your condition worsens, we will put you on the list, but not yet."

"When you are on the list and a kidney becomes available, it is offered to the patient most suitable to receive it, not the person at the top of the list."

"I have read up on CAPD. It seems a tiresome process and not without its own risks," said Richard.

"You are correct. It is a slow process and infection

is the biggest risk. Cleanliness of your catheter and equipment will be paramount."

"How often do I need to do this?"

"We will start your treatment three times a day," said Mr Cunningham. "It might be necessary to increase to four times a day in due course. Do you have any questions?"

Richard and his mother, Joyce, had many. None of Mr Cunningham's answers made them feel any better. CAPD was a prelude to a possible kidney transplant. It was also a stepping stone on his way to full haemodialysis. The consultant confirmed that he was unsure how soon Richard might need to start this treatment. He told Richard that he would spend two to two and a half hours daily performing the treatment. The doctor's primary concern was whether Joyce had enough room at home to store the complete bags and other equipment.

"How much space will all this equipment need?"

"It usually fills a cupboard in your home."

"What if we want to go on holiday?"

"You will need to take enough to cover the period you are away."

"What with that and our luggage, I'm not sure our car is big enough."

"Long-haul flights are out of the question I'm afraid," said the doctor. Eventually, the consultation ended, and Richard had to prepare to have his CAPD catheter inserted in due course.

Richard knew that without dialysis, he would face a slow death, so he saw it as the only option.

After the catheter was inserted, Richard had to sit

in a chair three times daily for his treatment. Each treatment lasted for about forty minutes. The first task was to drain the fluid from his abdomen. Then he detached the bag and sealed it. He put a new bag of fluid on a hook above his head. He cleaned the catheter and connected the replacement bag to it. He sat there while the fluid from this bag slowly drained into his abdomen. For two hours each day, Richard was awake but unable to do anything but sit and wait. He read quite a lot to keep himself from swearing too much. He read numerous books on almost any subject he could get from his local library.

As a teenager, Richard's hormones were kicking in. He would take frequent breaks from reading to eye the young girls who passed by his front window. When he was fit enough to go to work, he enjoyed looking at and talking to girls. He lived in the hope of losing his virginity. However, a catheter sticking out of his abdomen would forever be a blight on that hope. He liked women. They were very different, and he liked them different.

In line with previous advice, Richard's CAPD was increased to four times daily. His employer was very supportive and gave Richard space in the company medical office for some of the supplies he used during his midday treatment. He did the other three at home. One on rising, one as soon as he got home and the fourth as late as possible in the evening before he went to bed. Richard was placed on the transplant register in 2012.

On 12 February 2013, Richard received Sophie's second kidney.

10

Simone Jessop

Simone was born in Chiswick in 1998. Once a humble fishing village on the Thames, this area is now a wealthy hotspot favoured by celebrities. It's a leafy suburb with a charming village feel, unlike the industrial concrete, glass and metal of the rest of London.

Simone's parents would not be called celebrities, but they were affluent. Her father had a senior post in a City of London bank. He regularly took home significant bonus packages from his employer. His wife did not need to work and spent her time on the local coffee circuit or doing good deeds for the less privileged.

Simone liked Chiswick as it reminded her of the countryside. It still had large commercial buildings dotted around, but there was plenty of open space and the nearby river Thames. It was a small oasis of the countryside, captured by the sprawling concrete jungle that was London.

Simone would often wander down to the river Thames. She would sit by its banks reading. Sitting there on a warm afternoon, Simone became engrossed in her books. She did not hear any traffic noise or the roar of the aircraft landing at Heathrow or the City Airport. Caught up by the stories she was reading, extraneous noises were silenced.

She started her schooling at the Belmont Primary School before moving to the King Edward V1 High School for girls for her secondary education.

Simone was a bright child whose life was notable by an early diagnosis of Type 1 Diabetes. It was not a surprise that Simone would be diagnosed with the disease as her mother also had it. It was one of those diseases that could be hereditary.

Teachers at both of Simone's schools were well aware of her medical condition. Simone was not the only student at the King Edward V1 school suffering from type one diabetes. Romily Jeffries also had the disease. The two became friends as a result of their shared medical problems. They spent many hours supporting each other, comparing notes, and learning as much as they could about the disease and how to control it. In addition, they frequently spent time in each other's homes and developed a close bond that was more like sisters than just friends. They often slept over at the other's house.

Simone's physical appearance changed as she transitioned from childhood to adolescence. Her blonde hair grew long, and she was blossoming into a beautiful young woman. Despite her medical condition, she was upbeat. Her parents were protective of her and not just because of her condition. The streets of London were dangerous. They used her medical condition to convince her to stay safe, as they put it.

Simone, like the majority of teenagers, was very adept with social media. She had a small group of close friends, mostly female, but there were also a few males her age. Like most young people of her age, Simone kept in touch with her friends using social media. She had a separate account dedicated to managing her type one diabetes, and she also used it to stay in touch with others who were undergoing treatment for the same ailment.

She had the mySugr app on her mobile telephone. With it, she could connect it to devices, like compatible blood glucose monitors and health trackers. This also enabled her to automatically upload or manually add data. It included tags such as needle change, stress, work and alcohol, which can help her keep track of what might be affecting glucose levels.

There was also a Bolus Calculator, which calculated how much rapid-acting insulin she needed for mealtimes or corrections and was suitable if she injected her insulin or used a pump.

Simone, like many teenagers, was concerned about her appearance. She had natural blonde hair and was five feet three inches tall. Although she was not entirely unhappy with her body image, she often thought she could improve it. The unrealistic portrayal of the ideal female body shape on social media also contributed to her self-doubt.

She was aware of the latest fashion trends among teenagers but wore what she liked, not what was fashionable. She had a good sense of what colours and styles mixed and matched to make her look 'cool'. This included a range of makeup that was lightly and skilfully applied.

Her music tastes included artists such as Katy Perry, Adele and Ariana Grande. As she grew older, her music tastes changed with the fashion. As she matured, she liked to party but was very conscious of her intake of alcohol and her level of exercise. She was aware of how these things might affect her blood sugar. Simone was what could be described as a well-rounded young woman.

She had her own structured diabetic education programme that included information on dietary management. She was also aware that she would experience higher blood glucose levels in the days before a period occurs. This is because the effects of insulin are reduced at this time. Female periods were not just inconvenient and painful as they could also be very dangerous to her health.

Simone set period reminders on the mySugr app. She also used its calendar function to keep a checklist of her supplies. She did suffer from bouts of depression, but she and her friend Romily talked about being different and helping each other. They also used the school guidance counsellor.

Adding to Simone's anxiety was the fact that her mother died as a result of a severe hypoglycaemic episode when Simone was just fourteen years old.

Her devastated father, a city banker, hired a minder for Simone. He was out at work all day, and his daughter was vulnerable because of her medical condition. He wanted someone available between her return from home from school and his return from work. He hired a part-time retired nurse who also acted as a home help managing meals and diets for Simone. Gracie Wilson became an integral part of the family.

Despite Simone's diligent management of her condition and Gracie's help, her medical condition continued to worsen. She had several severe episodes of dangerously low blood sugar levels that happened without warning. They were not controlled with insulin. It was at these times when Gerald Jessop was glad that he had hired Gracie.

After another one of these dangerous episodes, and knowing her family history, her consultant, Mr Giuseppe De Silva, discussed the possibility of a pancreas transplant with Simone and her father.

Simone accepted it was an option worth pursuing and agreed to the lengthy testing process. When the full range of tests was complete, Mr Da Silva informed Simone that she was healthy enough to have a transplant, and it was now the best option, considering her age.

Many patients with uncontrolled insulin-treated diabetes have a combined pancreas and kidney transplant. Simone was awaiting a pancreas on its own. She was excited by the prospect of being free from insulin injections. She had anxiety because it was a significant operation, and she had been advised that there were risks of early complications. Mr Da Silva outlined the risks and how common or uncommon they were.

The figures he gave for the most common risks were what Simone and her parents considered acceptable.

There was a significant risk that she might need a blood transfusion within the first week of surgery. Simone would likely still be in hospital at this point in her recovery. She would get the transfusion swiftly if needed. She was happy to go ahead.

In 2012, Simone was placed on the transplant list.

On 12 February 2013, Simone received Sophie's pancreas.

11

The Call

In the early hours of a dark, cold, and wet February night, nine people received telephone calls that woke them from sleep. For eight of the homes, this was a call that would change one of the occupant's lives. The ninth home would not know until it was too late that she had received a call.

Roberta Wilson was one of the new 'like' generation. They were identified by their obsession with social media and the constant use of the word like, where it had no meaning. It conveyed membership in a particular youth culture. Older people used 'um' and 'ah' when stalling for time during a conversation. Youngsters used the word like.

Roberta had used the time she was performing her routine CAPD to update information on her several social media profiles. She also found time to search the internet for gossip on her favourite music stars and internet influencers. Roberta's life was governed by her addiction to all aspects of social media. She would visit friends, and they would spend time tapping away on their mobile telephone. Roberta, like many others, was addicted to her smartphone. Irrespective of where or what she was doing, Roberta instantly reached for her phone the second it pinged. She was never at a loose end if her phone was handy.

Roberta had been on her mobile telephone most of the day. She had fallen asleep without remembering to charge her phone. As a result, when the transplant coordinator called her, she got no answer as the battery in Roberta's mobile telephone was dead. Roberta got the bad news the following morning, at about the same time Rose Prentice was in the recovery room, having received the kidney that would have been offered to Roberta.

This was a fatal blow to Roberta as she would die before another suitable kidney could be offered.

Those transplant coordinators who got through to the other eight homes advised one of them that an organ that was a suitable match for them had become available.

All eight were asked if there were any new medical issues their hospital did not know about. All confirmed that they had no new medical conditions. Not even a cold between them.

The eight individuals were instructed to get to their transplant hospital with all due speed. They were reminded not to eat or drink anything before arriving, or they may make themselves unable to have the transplant.

In all eight households, the level of activity and anticipation was high. They had been waiting for this, some for many months. One of the tasks for those with jobs was to leave a message for their employer about their short-notice leave of absence.

For three of those called, it was a chance to delay their impending premature death. For the others, it would be a chance to lead a more normal life, liberated from the lifestyle limitations of their deteriorating medical conditions.

Some of the eight people had prepared for this moment by packing their bags with their pyjamas, dressing gowns, slippers, washing and shaving equipment, teeth cleaning tools, and regular medication, similar to how pregnant women pack for the hospital. Others were less organised and had to scramble around to find a bag and the items they needed. Upon checking her bag, eighteen-year-old Simone Jessop could not recall why she had packed tampons and sanitary towels.

Mixed in with the joy were feelings of anxiety and fear. All eight were going to have surgical procedures under general anaesthetic. For seven, those procedures would last several hours and up to ten in one case.

They had all discussed the benefits and dangers of their operations with their medical teams on several occasions.

A significant benefit was the expectation that transplant patients tended to live longer. Most patients would enjoy a better quality of life, and there might be fewer restrictions on the individual's diet. Work and travel might become less challenging. All reasons to look forward to the operation.

The risks were that seven of the operations were considered significant surgery, and that might lead to bleeding. There is also a higher risk of infection because of the medicines required to lower the immune system. Finally, there is no cast-iron guarantee that the new organ will work. The operation may even have grave or life-threatening complications.

Without the transplant, for seven of the patients, there was the certainty that their life could end prematurely and painfully. For the patient having Sophie's cornea, he might become blind in one eye. All eight thought that a reasonable chance of improved life was better than no chance, risks or not.

Arthur Morgan's cornea operation would be the shortest, at about an hour. Megan Thomas would expect to be operated on for up to ten hours. None looked forward to the operation, only a successful outcome.

Another emotion was summed up by Jacob Marshall as he rode with his mother to Harefield Hospital.

"I'm going to get a chance to live because someone died tonight," he said.

"Yes, you are," said his mother. "You should not think like that."

"I can't help it. We are happy while another family is devasted by loss. Hearing that their loved one is dead is bad enough, and then they are asked to make a decision about the removal of their organs. I can't imagine how that family must feel."

"Perhaps you could write to them to thank them when you get better."

"I think that's a good idea."

It was still a foul night as the eight started their

journey to their respective hospitals.

"Take it easy on these wet and slippery roads," Abigail cautioned her husband. "I may be getting an organ from someone who died tonight in a crash on roads like this." She had no way of knowing how accurate her words were.

All patients except Jack Gibson had a family to drive them to the hospital. Jack called the ambulance service authorised by King's College Hospital. They arrived at his home within minutes and whisked him straight to the hospital.

Arthur's wife drove him to the hospital. While he was collecting his bag and medications, she went next door to ask their neighbour to take care of their two children for the day. This had always been the Morgan plan. Luckily, the neighbour was at home and available to help.

It is difficult for transplant patients to live with the idea that some stranger with a generous disposition died to give them life. They often undergo survivor guilt. If they did not have that feeling this evening, they would soon experience it at some point in their recovery.

Although transplantation gives patients a new chance to live, the process entails an emotional roller coaster. Even after the psychological torture of the waiting period and the joy following a successful transplantation, various postoperative problems can cause physical and mental difficulties for patients.

There are significant risks of taking on transplant surgery, followed by the fear of acute or chronic rejection of the organ. There would be a lifetime of strict adherence to medication and lifestyle regimens. They all have the lingering side effects of immunosuppressive and other necessary drugs for their recovery. They may also have to adapt their lifestyles to maintain the new organ. Eight people were about to head into uncharted territory. A hopeful but uncertain journey.

12

The Transplant Centre

After the eight potential recipients of Sophie's organs were informed, they began to sort out their emotions and pack their bags, eager to get to their hospital.

Large teams of doctors, nurses, and technicians were on standby in eight different transplant centres. They were getting ready to receive the organs and patients for their transplant operations. Three of the hospitals were located outside London. These were the Churchill Hospital in Oxford, Addenbrooke's Hospital in Cambridgeshire, and Harefield Hospital in Hertfordshire.

The five in London were The Royal Free Hospital, King's College Hospital, St George's Hospital, Guy's Hospital and Moorfields Eye Hospital.

Eight organs had been harvested from Sophie, and this would lead to seven different types of transplant operations. The two kidneys would not go into the same patient. While there were seven different types of transplant surgery, the administrative procedures for each were similar.

The recipients had all been contacted and told that an organ that was suitable for them was available. They had been instructed to get to their designated hospital as quickly as possible and not consume anything before they came.

At the hospital, each recipient would go through the routine of having their identity checked when they met their respective medical teams.

The next step would be to take a recent medical history to identify any new issues that might make the patient unsuitable for the transplant. The telephone team at Bristol had already asked this question. The patients would quickly get used to multiple team members asking the same questions.

The first step for the transplanting team was to take some routine tests from the recipients.

The most crucial test for all organ recipients is matching and compatibility testing. It is actually three tests. The first test is to type the recipient's blood group. This should be known, but no harm in checking again. The second is a test of the blood group of the donor. As in blood donation, certain blood groups can only be given to a compatible blood group. For example, Blood type O is compatible with all other groups. Those with group AB are only compatible with another AB donor. Finally, there is the HLA test. This is also called tissue typing. This is the human antigen test. Its main objective is to identify the quantity of antibodies in the recipient's blood. If the number of antibodies is high, the chances of organ rejection can go up. The risk of rejection becomes more significant if there is a notable mismatch between the patient's body and the organ.

Transplant recipients are routinely asked about their current vaccine status. The reason for this is simply that unvaccinated patients are at a greater risk of infection and may be denied the organ, especially if their antigen levels are high.

This is because, during the transplant process, they will be given high doses of immunosuppressant drugs, which will weaken their immune system and make them more vulnerable to infections. Easily treatable illnesses like influenza and pneumococcal pneumonia can be fatal to an immunosuppressed person.

What happened next varied by the nature of the transplant organ, individual consultant preferences and hospital protocol.

The number of tests in each centre was governed by the organ that was being transplanted and the hospital transplant protocol.

Some, like Arthur Morgan, had few tests done. Others had a battery of tests to ensure they were fit to receive the organ.

The transplant team also take time with the recipient to discuss the likely risks and benefits of the operation again. The recipient has to agree to accept the organ after they have been fully informed of all the benefits and risks.

The transplant teams would be given any significant medical information relating to the donor's health that might be carried by their new organ. This was another risk that had to be explained and accepted by the patient.

This collective gluteal muscle covering is necessary, just in case something goes wrong and the patient is not informed that it might have been possible. No doctor wants the cost of their medical indemnity insurance to rise because of litigation caused by not telling a patient that some part of the process could go wrong. Some doctors quoted the percentage likelihood of each possible problem. It is unclear if patients awaiting surgery take in these statistics. However, it is reassuring to medical insurance providers that doctors warn their patients.

Personal information about the donor is not discussed beyond relevant medical details that would impact the operation. Not even the surgical team have that information.

If the potential recipient demands to know the ethnicity or faith or if the donor committed suicide as a reason for their possible refusal, they will not be told because the doctors will not know. If they were to be informed, the recipient might refuse the organ on ethical or religious grounds.

If Jack Gibson, a suspected racist, considered this question when the organ was offered, he did not express it. It may have been that staying alive was more important to him than his racial prejudice.

After accepting the offer, the patient would be visited by an anaesthetist. The anaesthetist checks the patient's fitness to undergo the operation. They explain the need for the various tubes that might be inserted into the body. They also explain how they manage pain during and after the operation to keep the patient comfortable.

Medical staff may ask patients a series of questions

to prepare them for the operation, which can make patients feel like they are being interrogated. This is done to ensure that no important details are missed, which could cause problems afterwards. All parties involved are dedicated to ensuring tasks are carried out meticulously, leaving no room for error or oversight.

Finally, the donated organ/tissue is checked to see if it is still suitable for transplant. If it is, the operation proceeds.

The length of the operation depends on the organ or tissue being transplanted and any medical issues or complications during the operation. The time these preliminary processes take before surgery depends on the required tests, with faster being better.

King's College Hospital was likely to have the lengthiest operation as it was the small bowel that was to be transplanted into Megan Thomas. It is called the small bowel because it is only about one inch in diameter. However, it can be up to twenty-two feet in length. The large bowel is three inches in diameter but is only six feet long. The average time for this operation is between eight to ten hours. Handling this exceptionally long organ while trying to place it into the abdomen of the patient is a very tricky task.

Hospital staff are reminded in many pamphlets to be mindful of patient dignity. Quite how any patient can look dignified in a shabby blue hospital gown with their bare backside hanging out is open to conjecture. Still, when the patient is under anaesthetic, they have no idea how they look. Dignity is not something they are in a state to consider.

Once all these procedures are complete, the patient is wheeled into the theatre preparation room. This is the area where the anaesthetic team will check that they have the correct patient and confirm the operation before putting them to sleep. This is the last thing any patient will see until after the operation unless they wake up while it is in progress.

13

To Sleep, Perchance to Dream

Anaesthetists tell their patients that they are going to put them to sleep. Veterinarians put animals to sleep. The two procedures are not dissimilar, except for the end result. The animal dies. If the anaesthetists have done their job right, their patient does not. The difference is that the vet administers an overdose of the lethal drug to render the animal unconscious before it eventually kills it. The hospital anaesthetist administers a therapeutic dose of the fatal drug to their human patients, taking them to a depth of unconsciousness just short of death.

Anaesthetic sleep is not the same as the sleep we get when we go to bed at night. If you are tucked up in bed asleep, and someone sticks a sharp blade into your skin, the pain will wake you up. When a surgeon does that while you are under anaesthetic 'sleep', you do not wake up. You feel no pain. Once under your skin, the surgeon can ferret around the tightly packed anatomy of your body. They can move muscles, blood vessels, nerves and even whole organs around without you feeling a thing. They can cut, remove, sew, saw through bone, add mechanical aids and replace organs without you waking up or feeling them there.

Jacob Marshall might be surprised to see the shape of the sternal saw used to split his breastbone. It was not unlike the shape of the Black and Decker drill he had in his shed at home.

Anaesthetic sleep is not really sleep as we understand that word. It is quite a different experience. It is much more like a coma than a bit of nightly shut-eye.

Medical textbooks describe general anaesthesia as a procedure that involves inducing a state of precisely controlled unconsciousness in a patient during surgery. The drugs used in this process can be very potent and may cause serious side effects, including death.

These potent drugs are in use thousands of times every day in hospitals all over the world. Patients might be forgiven for not knowing that the medical profession still does not fully understand how general anaesthetics work.

They know they block the nerves from passing signals to the brain, but how that happens is still a mystery. Blocking signals from reaching the brain prevents the patient from experiencing any pain or discomfort throughout the surgical procedure. This is good. Gone are the days when conscious patients had to consume a lot of rough spirits, bite down on a mouth gag, and be restrained by a posse of strong men during an operation.

Despite using these sophisticated drugs every day and not understanding exactly how they work, giving a general anaesthetic is almost an act of faith. There is much about the workings of the human body that scientists and doctors still do not fully understand.

The drugs used in general anaesthesia can also

adversely affect some of the organs in the body. A general anaesthetic is the medical profession's way of taking their patients close to death without actually killing them. It's a highly skilled operation.

In most operations, administering a single dose of an anaesthetic drug is typically not enough to maintain the desired level of sedation. The number of repeated doses depends on the drug used, the age of the patient, and the length of time the patients need to be unconscious. It is not uncommon for anaesthetists to use multiple drugs, typically the ones they are most comfortable with.

After the operation, it might be necessary to inject other drugs to counteract the effects of the anaesthetic.

Once the patient falls asleep, a tube is placed into their mouth. This is called tracheal intubation and is needed to keep oxygen flowing to the lungs. The most common side effect of intubation is a sore throat when the patient wakes up. Before the procedure, false teeth are removed, and the location of any crowned teeth is recorded. Wearing a face mask might be difficult if the patient has facial hair, as the mask may not fit properly. There is a slight possibility that a patient may become conscious during the operation.

A wide range of expensive medical equipment would be used on all eight patients. Jacob Marshall would have his heart removed and Abigail Gould her lungs. Both bodies needed to be kept alive while the new organs were implanted. Jacob and Abigail would be kept alive by heart-lung machines while the exchange of organs happened.

The heart-lung machine is a large device that takes over the functions of the heart and lungs to oxygenate and circulate blood during open-heart and lung surgery. The surgeons will work quickly to ensure that neither patient would spend more than three hours on the machine. After this time on the bypass machine, the already high risks to the patients increase.

During periods of high concentration in a complex surgical operation, the last thing any surgeon needs is a blown fuse in a machine or a power failure. There is no time to hang around and wait for an electrician to fix the problem. The show must go on.

A lack of electrical power to the machine would require it to be manually cranked to keep it operating and the patient alive. It would not look good on a death certificate if the cause of death was recorded as a blown thirteen-amp fuse on the heart-lung machine or the nurse's arm getting tired.

Humanity will no doubt work toward a future where the heart-lung machine will match the size of the human organs and can draw power from any nearby source. In 2013, Jacob and Abigail made do with the current model.

The eight operations all started at different times that morning. The duration of the anaesthesia varied considerably among the patients, with Arthur Morgan's being the shortest, just over an hour, and Megan Thomas's being the longest. Her operation would take nearly ten hours. As a result, the effects of the general anaesthetic would wear off quicker for Arthur than

Megan, who would suffer the ill effects for several days.

The duration of each operation was determined by the many steps needed to surgically remove the diseased organ before replacing it with a new one. Diseased kidneys are rarely removed. The new one is implanted into the patient's abdomen. It works perfectly well from there.

Once everyone in the operating theatre is ready and the patient is asleep, then the operation can proceed. All of the surgeons were very experienced, and all had a plan of campaign burned into their brains.

Start the first incision here, then do this, followed by that, and so on. All sorts of complications could arise after the first incision, throwing the estimated schedule and plan off course.

Napoleon Bonaparte reportedly said that no plan survives first contact with the enemy.

When a surgical operation is performed, the human body has no defensive plan to attack the invading surgeon. However, the immune system is designed to attack foreign objects such as implanted organs.

Once inside the body, the complex mass of bones, tissue, blood vessels, and nerves can be difficult to navigate. A tiny slip here or there could have devastating or even fatal consequences. This is not the time for the consultants to leave the theatre to relieve themselves.

Concentration throughout the complex transplant operation is vital for the whole team. The longer the operation, the more tiredness comes into play and the greater the chance of error.

As luck would have it, six operations mostly went

according to plan. Megan Thomas's operation did take longer as the intestine to be removed had already had several procedures before her transplant. Working around her scar tissue took longer than expected.

Rose's operation also took longer, as her vital blood and breathing functions fluctuated during the operation and needed to be stabilised before the team could proceed. Eventually, the first phase of all eight procedures was completed successfully.

Recovery from general anaesthesia can take time and varies depending on factors such as operation length, patient weight, and drug excretion rate.

After the operations, small parts of Sophie Coleman lived inside eight complete strangers.

14

Sweet Dreams?

When the human brain is under general anaesthesia, it is disconnected from the external world. It can generate an entire world of conscious experiences on its own. In effect, dreaming whilst under anaesthetic is similar to dreaming while asleep. Not all patients report having dreams while under anaesthetic, but many do.

The contents of their dreams seem to fall broadly into the four categories. Top of the list of dreams are those concerning loved ones and family. Next come dreams of holidays, followed by work and occasional erotica. Some of the eight organ recipients would report having dreams while under the anaesthetic, but not all of them.

After she regained consciousness, Abigail Watson thought she had dreamed about her first job. She remembered standing behind the cash register at the Woolworth store in Norwich, arguing with a customer about the pick and mix he had presented for payment. Abigail was sure he had eaten some before he got to the counter to pay. They had an argument before she relented and let him off. She couldn't recall if this had happened, but it played out in vivid colour and surround sound.

She spent seven days in the ICU after her operation. During those days, she had multiple episodes of dreams, hallucinations and nightmares. She also experienced nausea but did not vomit. Abigail found the experience in the ICU to be both frightening and hot.

Jack Gibson also woke up in the ICU, unable to speak because of the tubes in his throat. He had been dreaming about killing a man who had insulted his manhood. He thought the man was a black drug dealer in London but was not sure. "Anyway, the bastard's dead now," dreamed Jack, who was standing over a body in his dream.

Like Abigail, he had further dreams and nightmares while still in the ICU. Jack would also describe how he was afraid to sleep just in case he might pull out his tubes.

After waking up in ICU, Rose Prentice started to tell the person beside her bed about a holiday she and her family had to Southend when she was a child. She remembered this holiday because it was one of the few that she had. Rose had no idea who the person standing beside her was. The ICU nurse had heard it all before.

Neither Megan Thomas nor Arthur Morgan reported any dreaming during the operation.

Jacob Marshall came around in the hospital ICU dreaming about a beautiful girl he was in love with. He could not remember if she was real or an image he had created. Jacob knew she was the most beautiful woman he had ever seen. He wanted to be with her, whoever she was. For those few precious moments, it removed him from the horror of where he was.

When Jacob recognised where he was, he was pleasantly surprised to feel better than before surgery. He felt nowhere near as weak as he had done.

Simone Jessop dreamed about sitting in her bedroom with her friend Romily Jeffries after school, listening to their favourite music and chatting.

Richard Ainsworth reported dreaming, and his experience was the most detailed. It stopped when he finally realised where he was.

"Mr Ainsworth, Mr Ainsworth, Richard, you're back with us now. You've been having a bit of a bad dream. Everything is OK."

"I was having a bad dream," mumbled Richard.

"I know," said the voice beside him. "Everything is fine."

"What?" The hoarse question came from a man, neither awake nor asleep, who now had a sore throat.

"The operation went well. You're in the recovery room now. I'll just do some routine observations," said the voice.

"What?" Richard forced his heavy eyelids open. "What?"

"You will be fine. Everything went well. I'm just going to take your observations."

"OH, OK." Richard was still half awake and not in total control of his thoughts.

"I'll get you a drop of water in a minute," said the

voice that Richard could now see was attached to an out-of-focus blue blob beside him. He felt something on his shoulder, perhaps a hand. "You'll be OK in a few minutes. Just lie still."

"OK." Richard might have said OK if she had said she was going to poke his eyes with a hot needle.

While she was doing her observations, Richard remembered his dream.

He had dreamed he was wide awake and lying on his back on a bed in a dark room. He saw two thin ribbons of light, one high and one low, indicating a closed door with lighting on the other side. The dim light revealed a shadowy figure sneaking towards his bed in the darkness.

He did not know why but knew he had to act to save himself and his family from this menace. He wanted to leap from the bed to defend his family from the intruder. His mind commanded his body to act, and yet it did not. The only moving parts of his body were his eyes and mouth. He tried to shout, but no sound came out. His eyes darted from side to side, expecting to see help or salvation, but none came. He was paralysed. No matter how hard he tried, neither his arms, legs or any other part of his body would obey his brain's efforts to move them.

The figure crept closer, and his anxiety increased exponentially. He and his family were about to die, and there was nothing he was able to do about it. He tried to scream a warning, but no sound emerged from his lips. His heart raced to the point where he thought it might explode. All he could think of was death at the hands of this silent figure.

Slowly, the darkness faded, and the menacing figure was lost in a swirling dark grey mist. Richard realised that his eyes were shut and the images that he had seen were just a nightmare. It took some effort for him to force his eyelids open momentarily. All he could see was a white flat wall full of tiny holes. He had no idea where he was, and he was disorientated. He was exceedingly tired, and his eyes slowly began to close again. He heard a soft, gentle voice say his name, but he ignored it. Sleep was what he wanted. He was sure he had drifted off to sleep, but the voice was closer, and he felt his arm being caressed.

Richard guessed the person beside him had interrupted his dream as she woke him. She spoke to him again.

"You'll be staying here for about an hour before we wheel you back to the ward."

"Ward? What? Yes, OK." Babbled Richard, still some way from understanding where he was and why. He licked his dry lips. "I had a bad dream."

"I know, I heard you shout."

"Sorry."

"Don't worry, it's all quite common. You're fine now."

"I feel a bit sick."

"Are you going to vomit?"

"I don't think so, but best be prepared."

The nurse already had a handy receptacle. Nausea and vomiting were a common side effect of general anaesthesia. Any patient who had a previous operation where they experienced nausea gets antiemetic medication.

Vomiting can cause pain and dehydration. Transplant

patients usually get adequate postoperative pain relief. Taking liquids by mouth is impossible with a tube in the throat. These patients have liquids intravenously to keep them hydrated.

Unluckily for Simone Jessop, young female non-smokers were most likely to get postoperative nausea and vomiting. She did, twice, following her minor bleed repair.

Many years of experience in attending post-operative patients in the theatre recovery suite meant she hardly heard what Richard said. He was like so many before. She and her colleagues would carefully monitor the patient's recovery to the point when they could be safely returned to the ward. Job done.

Before he returned to the ward, Richard explained his dream to the nurse, who was only half listening as she carried on with her tasks.

It was probably as well that most patients did not remember what they said, coming out of the effects of general anaesthetic. Staff looking after them had heard much of it before. They might have a laugh at some of the patient conversations in the tea room after their shift. Most were instantly forgettable, incoherent ramblings.

As the effects of the anaesthetic wore off, patients became more lucid. Almost the first thought that crossed their minds was that they were alive. Next comes the question of whether the operation was successful. They had no way of knowing how long the operation had taken or whether it had been successful. The nurse had told Richard it was successful, so that was a relief.

They were still alive, possibly with a new organ or the old one still working as best it could. It would take some time before their brain could fully process what was happening.

Richard slowly regained consciousness and tried to make sense of his surroundings.

Looking up, Richard recognised the white wall with small holes. It was the ceiling of the recovery room. If, as the voice said, the operation went well, he was the grateful recipient of a new working kidney. This was good news.

He started to remember the anxious wait the previous evening. He and his mother, Joyce, had received an urgent telephone call to come to the hospital, where there was a strong possibility of a donor kidney if compatibility tests were positive. They were, and Richard was made ready for the operation. The Ainsworth family now had high hopes for a more normal life for their young son.

Waking up from the anaesthetic affected each patient differently. Part of the process is which anaesthetic is used and how frequently it needs topping up during the operation.

Arthur Morgan woke and recovered quickly because his operation was the shortest of the eight.

15

When Will It Start Working?

Organ recipients often ask whether the newly transplanted organ will start working immediately. The answer is that it's not a definite yes or no. A transplant from a living donor typically functions faster than one from a deceased donor.

For instance, a kidney donated by a living donor usually begins working within a day after the surgery. On the other hand, kidneys from a deceased donor take longer to 'wake up' after being frozen. It could take several days for them to start working and up to a week for the new kidney to work correctly.

Transplanted lungs usually start to work within a few hours of the operation. Fully operational lungs can take longer depending on many factors. These factors include the recipient's age and ability to clear their new lungs of debris after the operation.

Some heart transplant patients need mechanical circulatory support that might last five or six days.

Liver transplants can take up to three weeks to start working correctly.

Pancreas transplants usually start working straight away.

Cornea transplant patients might have to wait only a few weeks or up to a year for the new eye to start working completely.

The operation with the broadest range of possibilities is the small bowel. Recovery takes much longer than for any other organ transplant.

How long the patients might need to wait for their new organ to start working without fault depends on many factors. How ill was the patient before the operation? How old is the patient? How well is the patient responding to medication, and whether they do everything required to give the new organ every chance to work?

What is clear is that the time taken for the new organ to start working is different for everyone, even those having the same type of organ transplant.

In some cases, the transplanted organ does not work at all. In a few more heart-breaking cases, the patient dies during the operation.

16

Preparing For Discharge

All eight patients had undergone general anaesthesia for at least the duration of their operation. Those going into the ICU were kept under anaesthesia until they arrived safely in the unit, and some even longer. None woke up during their operation, which was a blessing to all concerned, especially the patient. It does not bear thinking about if a patient wakes up during the operation only to see their heart, lungs or liver being removed.

Neither the anaesthetists nor surgeons wanted any of their patients under general anaesthetics longer than necessary. Two of the operations caused the surgeon and anaesthetist some concern. One was that of Rose Prentice. Her blood pressure rose higher than they would have liked, and they spent some time bringing it under control before proceeding with the surgery. Megan Thomas's operation was complicated by the amount of scar tissue on her bowel that Mr Doshi and his team had to deal with as they went.

Arthur spent just over one hour under anaesthetic, while Megan spent nearly ten hours under.

The patient's post-operative recovery location

depended on how closely they needed to be monitored. Richard and Arthur both went from the operating theatre to the theatre recovery room and then to the transplant ward. Their initial post-operative recovery was the most straightforward.

Rose had been expected to do the same. However, the problems during her operation meant she needed additional management, at least for one day, perhaps more. She spent one day in the ICU.

During the operation, Mrs Mason noted that Rose's blood pressure and lung function were not as good as she would require. She put her surgery on hold until Rose's condition improved enough to restart. It took the medical team about ten minutes to return the two issues to a point where the procedure could continue.

Abigail Watson, Jack Gibson, Megan Thomas, Jacob Marshall and Simone Jessop also all woke up in their hospital intensive care units. A necessary step in their recovery process. A far from pleasant experience.

While in the ICU, Abigail, Jack, Rose, Megan, Jacob, and Simone experienced varying levels of delirium and difficulty concentrating. While there, they all also had dreams, hallucinations, flashbacks and nightmares. These are all common experiences of patients who are in Intensive Care.

Arthur Morgan and Richard Ainsworth both returned to the transplant ward of their hospitals after a stay in the operating theatre recovery room.

Moving from the ICU to a ward was seen by each

patient as a significant step on the journey to going home. The most noticeable difference between wards and the ICU was the lack of mechanical equipment and nurses. For a couple of days after leaving the ICU, there is still the need to monitor the patients and the transplanted organ. Specialist staff visited regularly to monitor the transplanted organ and the physical recovery of the patient. Multiple factors can affect their recovery time, such as the complexity of the surgery, age, and mobility. Whether it is painful or not, patients need to get up and start moving around as soon as possible.

When each patient was able to be discharged depended on the depth of their understanding of their condition and their ability to look after themselves at home.

There is a set process for recovery and aftercare in all transplant patients. It depends on the transplanted organ and how well the patients cope with what is needed to get back on their feet.

Recovery times can depend on how ill the patient was before the operation, their age and their resolve to get home as soon as practical.

Richard and Rose both received one of Sophie's kidneys, but Rose's age and previous physical condition complicated her recovery.

Patients must understand the symptoms of rejection and signs of infection before they are ready to be discharged. This knowledge will help them catch potential issues early on and avoid disastrous outcomes. Additionally, depending on the type of organ transplant, patients may need to make some lifestyle changes to aid in their recovery.

Jacob Marshall woke up in the Critical Care Unit of Harefield Hospital a full day after his operation was completed. He lay on his bed surrounded by various instruments. A casual onlooker might have trouble identifying the presence of a human body.

He had a breathing tube in his mouth. He also had several intravenous drips running into a vein in his neck. Added to the chaotic picture were the four tubes exiting his chest, two on each side. There was a wire coming out of his chest that was connected to a pacemaker that regulated his new heart rate. His urinary output was through another tube, as was the blood draining from his wound. Finally, he was connected to several mechanical devices measuring every aspect of his bodily function and keeping his new heart beating. The amount of technology keeping him alive and his bodily functions operating was staggering.

Later that first day, Jacob was rushed back into the theatre because his team had discovered a tiny bleed that needed to be mended. All the work done to close up his chest was undone before the team could look for the source of the bleeding. Once repaired and final checks done, Jacob's sternum was wired up again, and his chest closed. It would not be the last alarm that Jacob would have before he left the hospital.

Arthur Morgan's cornea graft meant he could go home the following day. Arthur could have been treated and home the same day, but his operation started later than was intended as the theatre had been occupied with an emergency. It took time to make it ready for Arthur's transplant.

Richard would spend ten days in hospital before being discharged home. The various tubes in his veins and abdomen were removed after a couple of days. By the day of his discharge, Richard's kidney was working very well, if not yet at total capacity.

Jack Gibson spent four days in ICU before moving to the regular transplant ward. In the ICU, his ventilator was removed as he woke up, but he had a face mask to receive oxygen. He was visited by a physiotherapist who taught him breathing exercises and coughing, which were needed to prevent breathing problems and chest infections.

Because of his history, Jack was visited by the liver team's specialist addiction nurse. He did not say that he would rip Jack's liver out himself if Jack started drinking again, but Jack got the message loud and clear. He said he was no longer an addict.

"You have a liver transplant because you were addicted to alcohol. You will always be an addict, and you need to remember that," said the nurse in a very stern voice. "You may have disappointments from setbacks during your recovery and think that a drink will make it all better. It will not. One drink could very easily start you back down that path again. One liver transplant is all you are going to get because of your addiction. There will be no second chance."

Jack got the message, but whether he would remember it if and when he had a setback. Only time will tell. His team recognised that he would have recovery setbacks and would be extra vigilant when they occurred.

The usual hospital stay after a liver transplant is about two weeks. Jack stayed a bit longer as he was not physically strong enough, and he was a slow learner when it came to his medication and signs of rejection, infection and addiction.

Rose only spent one day in her ICU but was not able to be discharged for sixteen days. Her team had a concern that she might have developed a blood clot, but this proved false. Her delayed discharge was partly down to the need to get her blood pressure medication dose under control in concert with all of her other medications.

As the oldest recipient of one of Sophie's organs, Abigail Watson took longer to recover in the hospital than usual. Her first five days post-op were spent in the ICU. It took her longer than expected for the new lungs to start working, nearly a full day. Her first breath after the operation was very strained. The ventilator had been doing the work for her, and now she had to start to do it herself. It was hard work. Her age and previous inactivity did not make it any easier.

She had difficulty coughing to clear her lungs of mucus and bleeding because of the pain it caused. As a result, her lungs collapsed twice. Fortunately, they were able to be inflated again.

Megan spent seven days in her ICU before moving to the specialist transplant ward. She was fed through a tube until it was deemed safe to allow her to start eating normally. She had a stoma that allowed the team to monitor her volume of waste and enabled camera access to the organ to check for rejection.

Megan stayed in the hospital for eight weeks before being discharged. As she lived close to the hospital, Megan was discharged to her home rather than staying in the hospital flat to be closely monitored.

Simone Jessop spent only one day in the ICU before moving to the transplant ward. She would spend fourteen days in the hospital before her discharge home.

Simone's new pancreas started working within two hours of being transplanted. Unfortunately, Simone suffered a minor bleed that needed urgent repair. Fortunately for her, it was a minor inconvenience requiring another surgery. It did not necessitate the removal of the transplanted pancreas.

She started eating three days after her surgery, and the tubes in her body were removed in three days.

While Arthur Morgan was leaving the hospital the day after his operation, six of the other patients were still in ICU. Arthur would return for two follow-up appointments before most of them had been transferred to the usual transplant ward and had any of their tubes removed.

Recovery from any transplant surgery is never easy. A question that is uppermost in patient's minds is whether the new organ was working as well as it should.

Patients often experience some of their pre-operative symptoms for a short period after transplant. Immediately after the operation, the patient's activities are frequently also limited by pain, muscular weakness, fatigue and sleep disturbances. This can reduce their physical ability.

Secondly, there is the question of their quality of life. There is an assumption that their poor quality of life before the transplant will improve. Will they be able to participate in their usual activities, return to work and socialise with friends? They know that immunosuppressant drugs depress the body's ability to fight infection. This might leave them open to an increased risk from any number of diseases that are usually benign.

Getting back to work is a strong motivator for transplant patients. They often face significant expenses after their operations and know that returning to work is essential to escape the feeling of uselessness or social isolation.

After their transplant, the patient's pre-operative depression and anxiety often decrease or vanish for a time. Depending on the results of follow-up appointments and tests, these can return quite quickly. A lot of effort is put into psychological support before and after discharge.

17

After Discharge

Each patient spends as little time in the hospital as is essential for their immediate recovery. They needed to be starting their new lives as soon as practical. Hospitals want patients home as quickly as possible because they need the beds for others. Delays to patient discharge are known as bed blocking.

One of the difficulties after discharge is related to the need for Social Care. If the patient was fit enough to return home with some short-term help, they would be considered fit for discharge. Hospital bed blocking is caused by the scarcity of Social Care in the community. Getting several Care agencies working in tandem for discharged patients is often difficult.

Social Care is the responsibility of the relevant Local Authority. The issue for them was their ever-decreasing budgets as successive governments reduced the amount of public funds for their use.

Social Care providers are often private contractors. They are required to make a profit to pay shareholder dividends and corporation tax. These were added costs that would not be needed by a public organisation. Fortunately, the eight patients required very little help from their Local Authority. What was needed was quickly available.

For Jacob Marshall, getting a heart transplant meant that he traded one disease that he could no longer live with for a situation that he could live with, but one that came with many restrictions and requirements. The same was true for Abigail Gould and Jack Gibson.

Jacob spent his life on very high alert because he was concerned that he could pick up any bug from anywhere or any person. He had days where he had bouts of depression and anxiety. He accepted that this was part of his new life. It was better than death.

Jacob was closely monitored for any signs of rejection. He had frequent biopsies of small pieces of heart material taken as required.

One of the main risks after a heart transplant is infection. The patient's body is immunosuppressed, meaning any passing bug could do some considerable damage. The stainless steel wires holding the breastbone together can become infected and might have to be removed. This did not happen for Jacob. Four months after his operation, he was informed that the wires no longer served any helpful function. They would not be removed because that involved another operation. Nicola Stephens did not think that was necessary. The metal in his chest wires might create problems if Jacob ever went through an airport or Eurostar security gates.

He did, however, suffer from high blood pressure that required medication to lower it. More drugs, just what he needed. He was already taking immunosuppressants, antivirals, painkillers, blood thinners, cholesterol-lowering pills, and medication to protect his stomach from the irritation of all the other drugs he had to take.

One bit of advice that was most unwelcome to Jacob was to keep out of the sun. He liked sunbathing, but Nicola Stephens warned him that skin cancer was quite common in heart transplant patients. He must always wear factor 50 sunscreen when outside, was her advice.

He was aware that sometimes even the best medical science fails. His watchword was vigilance. If he had the slightest indication of something abnormal in his recovery, he would contact his medical team. They were as anxious as he was to ensure their work did not fail.

He also had some survivor guilt, knowing that someone had died to give him his new heart. He remembered the conversation he had with his mother on the way to the hospital to receive his new heart. He had decided to write to the donor's family when he felt ready. This did not happen as the donor's next of kin was a friend because she had no family left alive.

Abigail Watson's recovery was very hard for her, partly because of her age. Abigail's new lung started to function the day following her operation.

Out of the hospital, her anxiety levels were very high, particularly on the two occasions when her lungs collapsed and had to be inflated. She also had hospitalisations when it became too painful for her to clear her lungs through coughing. All the muck in her lungs had to be cleared out.

She had to be especially careful to take her medication at the correct intervals. Her frequent follow-up examinations required a battery of routine tests, including the lung function test that she hated when her GP first administered it before her transplant.

Adil Patel's advice to Abigail to take things easy for the first few weeks was well received. However, she would be expected to exercise regularly and never start smoking again.

"Smoking nearly killed you the last time," said Adil. "You won't be so lucky the next."

Adil and his team gave Abigail a long list of things to watch out for after discharge. Shortness of breath, high temperature, fluid retention, new chest pain and feeling hot and shivery were some warning signs.

Hard as it was for her, Abigail did not smoke again, and she started exercising and watching her diet and weight.

As an alcoholic, Jack Gibson was either going to die or live on with a liver transplant. He had two prime issues after his operation. Despite not drinking booze, Jack would continue to be an addict for the remainder of his life. Temptation was everywhere. Jack would have the twin problems of recovery from his operation and freely available alcohol.

Second, he had faced enemy fire while in the army, but now he faced an unseen terror, a new liver. Jack did not understand why he was afraid but knew he was.

Like all transplant patients, Jack's biggest concern was infection because of his reduced immune system. He was still overweight after the operation. He took steroid drugs, and this resulted in him having high blood pressure and cholesterol. Jack blamed his weight on the number of pills he was taking. The terrible cheap food, filled with processed meat and fat, which was all he could afford on benefits, was not to blame.

He did drink, but only non-alcoholic drinks. When he went out, he would have soft drinks in pubs because not many had non-alcoholic beer. He did not mind that so much as the non-alcoholic stuff tasted like piss, he said. Quite how he knew what piss tasted like, no one wanted to know.

His visits to AA meetings became more frequent when he became anxious. This was true early on in his recovery when he thought he might be tempted to start drinking again. However, he became complacent and missed meetings and medication.

Jack wondered if the dreams or memories he was having about violence towards his organ donor were some form of psychological retribution for his actions towards his wife and children. This helped stir up his guilty feelings about his past alcoholic life.

Of the eight transplant patients, Megan Thomas had the most significant risk of her new organ failing. The gut is not a sterile organ, and infection is a very high risk after transplant.

Because the gut interfaces between the body and the environment, they are very immune-reactive, making them prone to rejection after transplant. Pre-operative tests for antibodies are even more critical for bowel transplant patients. The lower the level of antibodies, the greater the chance of a successful graft.

As a young man, Richard Ainsworth had a lot to live for. His CAPD routine was life-limiting, even if it did keep him alive. Like most transplant patients, Richard had survivor guilt. His kidney started to work less than a week after the operation.

He had a period after discharge where he and his medical team struggled to get the balance of his medication right. Like all transplant patients, Richard had to take a plethora of drugs, such as antivirals immunosuppressants, and painkillers. After the difficulty of getting the balance of his drugs right, Richard was careful in his approach to taking them.

On the few occasions, Richard felt unwell, he immediately called his renal team. He was seen immediately, and fortunately, the problems were minor. These episodes did little to help his feelings of anxiety and fear that he would reject his kidney and would eventually be put on haemodialysis. This was an extremely unappetising prospect.

Rose Prentice took longer than Richard to be discharged from the hospital because there were minor complications during her operation. Her kidney also took longer to start working to total capacity because of the issues during her operation.

Her recovery after her discharge was not particularly complicated, but she had severe anxiety on many occasions. Her anxiety was not just related to the possibility that her body might reject the new organ. She was experiencing peculiar dreams or recollections of unfamiliar places.

Her team suggested that dreams about the organ donor were not uncommon. She was invited to talk to one of the team's therapists. This seemed to help, at least until the next dream happened.

One of the most significant issues for Simone Jessop after her transplant was hair loss. This occurred sometime after she left the hospital. For a beautiful young lady, this was a significant problem easily cured with minoxidil.

She had bouts of anxiety brought on by the hair loss and the odd memories she was having about schools she had never heard of. She ascribed this to having memories from the organ donor, but her medical team assured her this was not possible. They said that transplant patients often had dreams about their donor. It was all quite natural. Simone was not so sure but bowed to the professional experience.

She had been fastidious in the management of her diabetes, and she carried that on after her transplant. She was careful about catching infections and used face masks for much of the early months after discharge. She also had a ready-made excuse for not cleaning her room. The risk of hernia in a pancreas transplant patient is high. Heavy lifting is to be avoided. She did not tell her mother that ten kilos was the maximum advised weight for her to lift. Her father was sure that shoes and underwear were perfectly OK for his daughter to lift, transplant or not.

Arthur Morgan was out of the hospital the day following his operation. His time spent under general anaesthetic was the shortest as he only had a cornea grafted to one eye. The rest of his body was functioning as well as it had been the day he entered the hospital.

His eye might take up to a year to return to normal vision.

He would have to wear a patch over his eye for a few weeks. He heard the comment about forgetting his parrot more than once when walking in the street.

He also had to keep water out of his eye for at least a month. He bathed rather than showered during this period. He was fortunate not to have any of the other complications associated with a cornea transplant.

Sometime after his surgery, Arthur's new cornea reminded him of how poor his vision had become. As with most transplant patients, Arthur was daily reminded that someone had died to give him better sight. He would be forever grateful.

Many transplant patients suffer anxiety and other psychological disorders to a greater or lesser degree. In a significant number of these patients, anxiety is evident before the transplant takes place. All eight patients had some form of psychological evaluation before their operation. It was known that post-operative psychological problems could have a significant impact on patient compliance and life expectancy.

For some, the follow-up appointments were almost routine. For others, they would have to undergo a battery of tests. The doctors had to check the health and operation of the organ.

The primary purpose of these appointments was to check the medical health of the organ and the patient.

Typically, surgeons and their medical teams concentrate on the medical aspects of patient recovery. For Mr Allan Cunningham, that was about to change unexpectedly.

18

I Have A Dream

Allan Cunningham's research registrar, Mr Graham Bradbury, had been running his transplant clinics where he regularly met Richard Ainsworth. Richard's medical progress was as expected. There were some ups and some downs. However, about five months after his operation, Richard told Graham that he knew the gender of his donor. His kidney had come from a woman. Graham dismissed the comment. He knew that organ recipients often had dreams of their donor. Guessing that the donor was a woman had a fifty per cent chance of being correct.

At the next appointment, Richard said he knew that the lady donor's name was Sophie Beckinsale. Graham still did not take the revelation seriously. His patient could have picked that name from anywhere. It might even be an acquaintance of his.

The appointment following this, he started to relate stories about Sophie's time at university. He told Graham that Sophie had gone to De Montford in Leicester. The stories were detailed enough for Graham to wonder what was happening to his patient.

"Have you been to De Montford, Richard?"

"No. I never went to university. I never got good enough results at school."

"Have you visited Leicester where this university is situated."

"No. I have no idea what the place is like except from the memories or dreams that I am having. What does it mean? Are these the memories from my donor?"

"I don't think for one minute they are. Memories are stored in the brain, not organs in other parts of the body. I just can't see any way in which memories from one person can be transferred to another. The two people involved, either need to meet to discuss them, or talk, or some other form of written or video messaging. Do you know or have you met anyone called Sophie, what did you call her?"

"Beckinsale, and no, I've never met anyone of that name. I was wondering if I'm going bonkers?"

"You may be quite stressed by the whole transplant experience. It's quite normal when you get another person's organ to keep you alive. I don't think you are going bonkers, as you put it. I don't know what's causing this dreamlike situation. Perhaps it will fade as you become stronger mentally and physically."

"OK doc, I'll try to ignore it."

Graham did not know the gender or name of the donor. The only relevant information passed on to him and the surgical team was the state of the organ and any of the donor's medical history that might be relevant to the operation.

He initially put Richard's comments down to PTSD or fantasy, which he had told his patient was not unusual in transplant situations.

Richard started to give more detailed information about this fictitious donor's life at De Montford. Graham became concerned. This was not some hazy, transient dream, as far as his patient was concerned. It was real. Graham could not understand it. His patient seemed to be obsessing about the donor. He appeared to have constructed a significant amount of detail about one aspect of the life of this unknown donor. He gave Graham detailed information about bits of this woman's life at De Montford University in Leicester, a place he claimed he had never visited.

After this most recent episode, Graham decided to raise his concerns with his Consultant, Mr Allan Cunningham, when they next had a regular meeting.

Towards the end of the meeting, Graham indicated that he wanted Allan to see Richard Ainsworth as soon as convenient.

"What is your concern about this chap? His numbers look fine," said Allan.

"Yes, they are. It's not his medical condition, that worries me, it's his mental state."

"What about it?"

Graham then explained what Richard had told him about the donor.

"He initially said that he knew that it was a woman who donated his kidney. At the next appointment he said that her name was Sophie Beckinsale and she attended De Montford University. I don't remember knowing that his donor was a female. I asked him where he got his information and he told me that he had memories and flashbacks from her life."

"Did you just say that he had her memories?" Said an astonished Allan Cunningham.

"That's what he says, and he is quite adamant. He asked me if he was going bonkers, as he put it. I didn't think so, but I'm no expert. Was his donor a woman?"

"I think so, but I don't know her name, only that she was fit, and she had no medical problems that might affect the recipient. I think you are correct. I need to speak to this young man as soon as possible. We it might be wise to have a psychiatrist present at the appointment. Can you get him in ASAP, please?"

Ten days after this conversation, Allan Cunningham, Graham Bradbury and Dr Madeline Walsh met Richard in Allan's office.

Allan had already briefed Madeline and Graham on what he had learned. He had spoken to NHSBT in Bristol and discovered that the donor was a woman called Sophie Coleman, born Beckinsale. The information from Bristol confirming the donor's gender and name concerned Allan.

"I spoke with the team in Bristol. They gave me her name and assure me that no one else has asked for this information. I asked if someone might have accessed their system remotely. They have no indication or evidence that their IT system has been hacked," said Allan. "I don't know what to think, or where to start when this young man turns up."

"What if I take the lead?" Asked Madeline. "I've never met the man, so I can start fresh."

"I can't think of anything better," said Allan.

Madeline started the meeting with Richard by explaining what she was and why she was there.

"As you can imagine, Richard, what you are telling us here is very unusual. None of us have heard anything like it before. Can I start by asking you to explain to us how you have come by the information you gave to Mr Bradbury at your previous meetings?"

"The information is just here, in my head," said Richard, tapping his skull. "I seem to know a lot about this woman's university days, but I know nothing about any other aspect of her life. It's very strange."

"Can I act a little like a policeman and ask some questions that might help us understand?"

"I've nothing to hide. Fire away"

Madeline asked about De Montford University. Had Richard been there. He had not and had not even been to Leicester. The questions from Madeline continued, but Richard gave no clue as to how he had come by this information except to say he had memories and flashbacks.

"It's all in here," he repeated, tapping his head. "I've never been to Leicester. I've no friends or acquaintances called Sophie. Nor do I know any family called Beckinsale, except the one in my head."

After exhausting all their questions, the three medics looked at each other blankly. The young patient sitting before them seemed lucid and confident. He had not deviated in any way from his story.

"Since our last meeting, Mr Bradbury, I've remembered more about Sophie," said Richard.

"Have you, and what might that be?" Asked a

disbelieving Graham.

"She studied Business Administration at De Montford. I know that she liked to eat in the Kimberly Library Cafe just off Gateway Street, and the Riverside Cafe off Newarke Close in the Vijay Patel Building. It was there they used to hosts pop-up events that Sophie attended. She even got occasional tickets for Leicester Tigers Rugby games at Wellford Road. She liked to watch the muscular men in their tight shorts. She liked well-built strong men. She did not know much about the rules, she just like the big muscular men in shorts."

There was a stunned silence in the room. Eventually, Madeline spoke.

"Do you know why she liked these sort of men?"

"Best not ask as I can't say with a lady present."

"I understand. You have given us this information, but there is no way in which we can verify that it is true. You could have looked it up on the internet. Is that what you did?"

"No. You could speak to Joan Tapp, she was a friend of Sophie and lived in the same block of flats. She was in the room next to Sophie. They spent a lot of time together. Joan studied chemistry."

Another stunned silence.

"Thank you, Richard," said Allan Cunningham, who did not know what else to say. "We'll need to take our notes away and think about what has happened here today. We've never heard anything like this before."

"I guess you're all a bit sceptical," said Richard. "I would be too in your position. I have no idea how I came to know this stuff. It's just there, inside my head." He tapped his skull again. "I'd like to know where it came from if you ever find out."

"We'll let you know," said Allan, and the meeting concluded.

After Richard left the room, the three medics looked at each other, wondering where to start.

"You're the expert in this field, Madeline," said Allan. "What's your take on what we have heard?"

"I would be inclined to say that he is delusional, but he seems certain, and has challenged us to prove him wrong. I suppose we could follow up on his comments. The information about the eating venues is likely to be online somewhere. I know that Leicester Tigers play at Welford Road and if I know, then so might he, if he follows rugby. There are two unknowns. Did this donor study business administration at university? That should not be hard to find out. The mention of her friend, Joan Tapp could be a good point to disprove his so called memories. I'm sure the university can get us some information on that. They will have records of students and their courses."

"Apart from that, I'm at a loss. I think we would all agree that he could not have got an organ from this woman and it came with some of her memories. It's not possible. Still, he seems sincere."

"I can think of no way in which this young man could have acquired any memories from the donor. It's just not possible. Graham, have you seen any research or papers that might explain what we have heard here today?" Asked Allan

"Nothing in the medical journals or anywhere else that I have read. Mind you, I've not read the psychiatry journals or papers, that's more your field, Madeline," said Graham. "Like you, Allan, I have no idea how any part of a donor's memory could be lodged in one of their donated organs. It's just too bizarre for words."

"I'm with you two on this," said Madeline. "The young man seems sincere and not delusional, is my initial impression. He has given us some information that is unlikely to be readily available on line. Perhaps we can start to unravel this problem there. I'll leave it with you two for the time being. Keep me in the loop as he might need treatment."

Allan asked Graham to nominate someone in their team to contact De Montford University.

"They should be able to tell us if this donor did study business. They should also be able to confirm if they had a student called Joan Tapp who studied chemistry at the same time."

"I'll think about contacting the transplant team in Bristol when I get back. I'm at a transplant symposium in Birmingham this weekend. Let me know if anything happens. Once we have that, we can decide what to do next."

19

A Surprising Meeting

On the first evening of the conference, Allan met Adil Patel, an old colleague. After the usual pleasantries and idle chat over drinks, Allan began to tell Adil about his patient, Richard Ainsworth. Allan felt he needed to talk to someone, if for no other reason than it was a strange but absorbing story.

"I have this patient who had a kidney transplant in February. His operation and immediate recovery were unremarkable. What puzzles me about this patient is that several months after the operation, he started to tell my team that he had information about the donor."

"At first it was the gender. He said he knew his donor was female. For the next few appointments he related more detailed information including her name and her time studying business at De Montford University. We thought it was some form of delusion, so I called in my head of psychiatry to take part in one appointment. She was mystified because of the level of detail he seemed to have, and the certainty with which he described his so-called memories."

"Yes," said Adil. "The donor was Sophie Beckinsale."
"How on earth can you know that?"

"I have a patient called Abigail Gould who had a lung transplant in February. She also started having memories some months after her operation. She quoted the woman's name and seems to remember parts of this woman's life, just as your patient does. Like you, I could not believe it and was prepared to put it down to some aberration, until you told me your story. It's the same name. Have you checked the transplant register details?"

"I got the information yesterday. The donor was Sophie Coleman who was born Beckinsale in Smisby in Derbyshire."

"So, two patients who had organs from the same donor are having similar symptoms, if that's what you could call these flashbacks. Medically, that's not possible. We transplant physical organs, not metaphysical memories. Memories are contained in various parts of the brain and as far as I am aware, not in any other organ of the body. I can't believe that I transplanted an organ that contained part of the donors memory. It's just not possible. It must be some form of psychological disorder in both of our patients."

"If it is, it's strange that both of our patients who had an organ from the same donor have the same mental condition," said Allan, knocking back the remainder of his scotch. "I'll get us another."

When he returned to the table, Adil continued.

"This is far too strange. My patient lives in Bury St Edmunds in Suffolk. Where does your patient live?"

"Sutton in Surrey. They are not neighbours. Are they related do you know?"

"Abigail Watson was born Gould in Norwich. What's

your patients name?"

"Richard Ainsworth and he has lived all of his life with his parents in Sutton. If they are related, it's a distant branch of a family tree. I seem to remember him telling me that his relatives lived in the Cotswolds and the West Country, somewhere."

"So, we each have a patients that claims to have the memory from what appears to be the same donor. There seems to be no familial connection between the two patients, as far as we can tell at the moment. What on earth do we do now?"

"Damned if I know, Adil. I suppose it could make a good paper for a future symposium or conference."

"More likely to see us both locked up in Broadmoor as criminally insane."

"That's closer to the truth. Still, now that we both have the similar issue, we should do something to investigate."

"Where on earth would we start? No one in their right mind would give us time or money to investigate such a preposterous idea. We'd have to do it on our own time."

"Well, my patient has one of the donor's kidneys, yours has the lungs. There might be others who have received other parts of the body. Perhaps we might start there. We should get on to the transplant service and check who else has body parts from this donor. We can see who they might be, and if they have a similar issue."

"It's a long shot, Allan, but I suppose it's the only place to start."

"I wonder what we might find. Our patients might be the only ones with this ladies organs. If there are any other recipients they might not have the problem or all of them have it."

"We should keep this quiet for the time being. Keep the insanity between us. I don't want my Trust Board hearing about this. They might have me referred to Lara Kerr."

"Who is she?"

"My Trust's senior Consultant Psychiatrist. I get on quite well with her, but she would have trouble believing our story. I'm not sure I believe it. Still, it's comforting to know that there are two of us with this delusion."

"As I said, I've had our head of psychiatry, Madeline Walsh attend a meeting with my patient. She has seen and spoken to him. She is mystified. She thinks he might be delusional, but worried because he is so certain and has given us information to check that would not be easily found by an internet search. If it's a delusion, I hope it's not ours. We are only reporting what we have heard."

"Think about it, Allan. They'll say that our patients are having delusions, and hearing voices in their heads. The only piece of proof that we seem to have is that you have discovered the name of your recipients donor. I don't know that my patient got their organ from the same donor."

"Come off it Adil. You mentioned the donors name before I told you who my patient thought it was. It's easy enough to confirm with a call to the transplant service."

"I'm sorry, Allan. If two other people told us this story, what would you think?"

"You're right, Adil. I'd think they were crackers."

"I'll phone the transplant service and get the name of the donor of the lungs. I'll also try to get them to tell me how many other recipients there were, and their details."

"If they won't cooperate, what then?"

"I'll spin them a story about our meeting at this conference. I'll tell them we might have found a possible medical issue with our recipients that needs further investigation. If there were other recipients, we need the names of their consultants to check their patients. I'll ring you and we'll concoct a story before we contact them. No point in alarming them if their patients have no symptoms, or delusions."

After the Symposium, Adil got through to the transplant service and spun his story. He needed to know if there were any other recipients of organs from the donor. Adil related a largely believable story about his meeting at a conference with Allan Cunningham. The issue was that he believed their patients received organs from the same donor, and both patients reported some strange but similar side effects.

The transplant service supervisor confirmed to Adil that he was one of seven other recipients and added the names of the transplant consultant surgeons.

Adil already knew about Allan Cunningham. He learned that Mrs Isabel Mason had transplanted the second kidney. Mrs Nicola Stephens transplanted the heart, Mr Keith Isaac inserted the liver, Mr Giuseppe De Silva the pancreas, Mr Kasara Doshi the small bowel, and Professor Mohamed Khan did a cornea. Those were all of the harvested organs reused.

Adil telephoned Allan Cunningham with the news. "There were six other transplant from this same donor. I have the names of all of the surgeons." He listed their names. "The transplant service know that you and I have similar concerns about the donated organs. I described our situation as a minor medical issue that might need further investigation. I promised to keep them informed of our findings."

"That's good. What do you think we should do now? How do we approach the others?"

"Why don't we approach just one and see what we discover?" Said Adil. "I've met Keith Isaac. Why don't I contact him?"

"That's a good idea. I've met, Isabel Mason I'll call her. Whatever we discover from those two, I suggest we continue to call the others as a matter of course. We need to check all eight donations to discover if some of the others have the same side effect, or even all of them."

"Since we last spoke, I have thought about what we have found. I'm at a complete loss as to what possible anatomical mechanism could have triggered this side effect," said Adil. "I'm not even sure that it is a side effect rather than some mental aberration in our two patients."

"You'll get no insight from me as I'm in the same dark room. Let's just contact these two first and compare notes afterwards to decide what to do next." With that, the conversation concluded.

20

More Dreams?

The woman sitting at her office desk did not give the impression of being a doctor who cut people open and transplanted organs for a living. She stood five feet eight inches tall, with long blonde hair. She was slim built with dark green eyes. She was often seen around the Royal Free Hospital in London dressed elegantly, more like the CEO of a large multi-national company than a surgeon who dug around in human bodies for a living.

She had completed her rounds for the day. She was sitting in her office finishing up her notes on her patients when her telephone rang.

"Isabel Mason."

"Good afternoon, Isabel, it's Allan Cunningham."

"Good afternoon, Allan. To what do I owe the pleasure of your call?"

"It's very odd, Isabel. I'm not entirely sure how to start. Do you have time to spare?"

"I can make time for you Allan."

"Thanks. On the twelfth of February, I did a kidney transplant into a twenty-one year old man. I have been in touch with the transplant team in Bristol who tell me that you performed the same operation on the same day. Are they correct?"

"The twelfth of February this year, yes, I believe I did." Isabel did not hesitate when she recalled the date and the operation.

"It seems that we both transplanted organs from the same donor and that's why I'm calling."

"Is there a problem with the donor or the patient?"

"I'm not sure how to describe it, but I have an odd issue with my patient. I need to describe it and get your input, to see if you can help."

"I'll try."

"I recently attended a symposium in Birmingham where I met Adil Patel from Addenbrooke's. He's an old acquaintance. I was chatting to him about my patient's issue when he said that he had the same issue with a patient he had done a lung transplant on the same day. We discovered that his lung had come from the same donor as our kidneys."

"Are you telling me that both recipients from this donor had the same issue?"

"Yes."

"What is the issue?" Isabel did not seem concerned there might be an issue with the donor.

"It's really, really strange because it is not, as far as we can say at the moment, a purely medical issue. We're not sure if it might be a psychological issue. My hospital head of psychiatry thinks it might be, but even she is not sure."

"Sounds interesting, what's the problem?"

"Several months into their recovery, both of our patients started to report knowledge of the donor.

They started by saying that they knew the donor was a woman, which I have discovered was correct, when I called Bristol. Soon afterwards, they both identified her name and they started to report specific memories of the donor. The two patients did not have the same memories. They seemed to have knowledge of different parts of the donor's history. Almost as if the organs they received contained different segments of her memory, not all of it."

"So, what do they know about Sophie Coleman?"

"Oh my God. You as well?"

"It would seem so. My patient is a forty six year old woman who could be described as ordinary. A bit of a dreamer in reality, perhaps even a fantasist. She started by saying that she knew the gender of her donor. Like your patient, she went on to state her name and then memories of places she had never visited. She claimed these were the memories of the donor. We wondered if the trauma of her transplant had affected her mental capacity."

"Adil is speaking to Keith Isaac at King's College, who transplanted this donor's liver. If he reports the same phenomenon we have a puzzle. There is nothing in the literature about transplant patients actually getting donor memories or feelings. Some snake oil salesmen will try to convince gullible people that their remedy cures emotions associated with bodily organs. What trial papers there are, tend to set out to prove an hypothesis, rather than find the truth."

"I've never heard of anything like this before, have you? You seem to be taking this seriously," said Isabel.

"I am, up to a point. Both of our patients have given

us personal information that would not be found on an internet search or be in open source documents. I've got one of my team trying to verify if any of these bits of information are correct. If we find nothing, then it's obviously a psychological issue. However, if it can be verified, then I don't know what to do next. It might be an elaborate hoax."

"If it is a hoax, then it would have to have been devised after the operations. None of the recipients would have known that they were going to get an organ, let alone for the same donor. I'm not sure my patients would be able to maintain secrecy if it were a hoax. Are you checking the other recipients?"

"We had only planned to do that if you and Keith had the same issue. Adil will ring me with the results of his conversation. At that point we will have to make a decision as to what we do next."

"Just to add to your pool of information, my patient remembers visits to Arizona in the USA. It seems that's where the donors in-laws lived. She was a bit hazy at first but she has become more convincing as she seems to remember more. This is impossible, Allan. We both know it is. There must be a logical explanation."

"Until I spoke to you and Adil, I would probably have agreed. Now, I'm not sure that I am not paranoid. If we published this information, Adil suggested we might end up in Broadmoor."

"He might be correct and I would need to go with you both."

The conversation continued in the same sceptical way for about ten minutes without progress towards a rational solution to their problem. Eventually, realising they were getting nowhere, the conversation started to wind down.

"Adil and I discussed contacting all of the other consultants to see if they had the same issue. If, as I now suspect, we all have the same issue, then I would propose we arrange a meeting for all of us, in person or online, to discuss what we do with what we have."

"That seems like the sanest thing I've heard so far. With her approval, I'm going to tape my patient's conversations. I suggest we all do the same if we find our worst fears realised."

"Good idea. Thanks for your time, Isabel. I'll keep you updated as we go."

"Thanks, Allan. Your call changed a dull day into anything but routine."

A week after his call to Isabel Mason, Allan got a call from Adil Patel.

"Allan, I spoke with Keith Isaac at King's College two days ago. He transplanted our donor's liver into an ex-army man."

"And?"

"And his patient is having the same issue with memory or flashbacks that our two patients have. These seemed to have started around at four or five months post-op, the same time as those of our patients. Like ours, his patient began by reporting her gender, then he used her name. After that he began to divulge more detailed information about her abusive husband."

"His patient said that her husband was a drinker, a womaniser and could be violent towards her."

"Anything to suggest that this was memory or some

trick they are all playing on us?"

"Not as far as Keith can ascertain. His patient left the army after some experience of combat and became drinker himself. When he had a few drinks, he would come home and slap his wife around. She took their two children and left him as a result."

"Keith's says that his patient did not seem to be a womaniser because he told Keith that he was often incapable through drink. His excessive drinking eventually led to him getting the liver transplant."

"Until I spoke with him, Keith wondered if the man's combat experience had brought about a PTSD delusional episode. He initially wondered if his patient was projecting his own life experiences onto an unknown organ donor. Once his patient started using her name, and he knows that there are other similar cases, he's as baffled as we are. What about your call?"

"I have to report that Isabel Mason's kidney patient is having the same type of incident. Just like the others, her patient started detailing the gender, then the name followed by trips to Arizona. Her patient has never been out of the UK. It seems that was where the donor's husband hailed from, and his parents lived. Her memories of the in-laws street and town are quite vivid, as are some of the trips that the donor took with her in-laws in the USA."

"So, that's four of the eight who have the same issue. I cannot believe that this is possible," said a clearly bemused Adil.

"I suppose all that we can do now is to contact the other four surgeons to see what they have to tell us. Do you know any of the surgeons left on our list?" Asked

Allan

"I know of Nicola Stephens at Harefield. I met her once or twice over the years, but she probably won't remember me. I'll speak to her," said Adil.

"I don't know Da Silva or Khan, but they are based in London. I could always meet them both for a coffee somewhere together. That would leave you with Kasara Doshi at the Churchill, if that's OK?"

"I'm fine with that. I'll call you when I've complete my mission, M."

"Very amusing, Adil. We're not James Bonds conducting secret operations. I was about to say scientific investigation, but I'd hardly call it scientific. It's more likely to be some paranormal hocus pocus, mumbo jumbo. I'm wondering if we should order an Ouija board or call Ghostbusters. I'll hear from you as soon as you've talked to the others."

21

Even More?

Almost a month after their previous conversation, Adil and Allan had another.

"I met both Giuseppe De Silva and Mohamed Khan at coffee shops in London," said Allan.

"What was the outcome?"

"It pains me to say that they report the same as the others. Just like the four patients we already know about, both of their patients started by remembering the donor's gender at about four or five months post-op. As time passed, they remembered her name and then much more detailed information about a small part of the donor's life."

"Mohamed Khan's cornea transplant patient remembers Sophie's husband's death almost six years ago. Giuseppe's patient is a young woman who seems to remember Sophie Beckinsale's primary and secondary school days. They give the donor's name as either Beckinsale or Coleman depending on what period of the donor's life they claim to have her memories."

"How these claimed memories can have anything to do with the pancreas and cornea that the patients received is beyond me, like everything else about these patients. That makes six out of the eight with the same issue. What did you find?"

"The same for my two. Nicola Stephens heart transplant patient remembers Sophie meeting and courtship by her husband, and their subsequent marriage. Kasara Doshi's patient remembers Sophie's life after the death of her husband. There was a great deal of anxiety and guilt in her life after the death. She does not know why. Her memory seems to stop the day Sophie died, but she has no memory of her actual accident that caused her eventual death. The memory seems to stop during an argument in a hotel with someone she called Roberto Caruso. So far, she has not indicated who this Roberto Caruso is, or how Sophie Coleman knew him."

"Nicola Stephens patient is only twenty-four. When he first started recounting memories, Nicola wondered if he was thinking about a young woman in his life, but she discovered that he had no such attachment."

"It seems we have an impossible puzzle. All eight patients who received organs from Sophie Coleman also seem to have received a small part of her memory. I know what I just said, and my scientific, medical brain tells me that it is pure rubbish. It's not possible." Allan's voice contained some astonishment and scepticism as he spoke.

"There was a time when no one thought it possible that men could walk on the moon, and yet, they have," said Adil. "Maybe there is a rational explanation that we have not yet discovered."

"When I spoke with Isabel Mason and the other two, they were all very sceptical, but wanted to be kept in the loop. Did your conversations elicit the same response, Adil?"

"They did. I think they got some small measure of comfort from knowing that they were not alone with this impossible problem. Doshi and Stephens had involved their psychiatric specialists to assess their patients. Just like your psychiatric consultation, nothing useful seems to have come from either."

"It seems to me that the next step is to get all eight of us together to try to work out what we do next."

"I'm at a loss, so I'll go with that," said Adil. "We are all very experienced medical specialists, I'm sure we must be able to come up with some explanation of this isolated phenomenon."

"I have been making some notes as we have gone along," said Allan. "I'd like to sift through them to try to make some sense of what we have discovered. I need to get my head in order. Can I call you on Monday when I've had a good look?"

"I'm operating on Monday," said Adil. "Can you call me early evening when I'll be at home?"

"I've also got a list on Monday. I'll ring you on your mobile in the early evening, assuming all goes well with both our surgeries."

22

What Have We Learned?

"Good evening, Allan."

"Good evening, Adil."

"I assume your list did not run over."

"No, nor yours?"

"No, everything went as planned, which is not what can be said of this transplanted memory issue."

"Have you had time to look over your notes and make any sense of what has happened?"

"Look over my notes, yes. Make sense of it, no."

"What now?"

"I have tried to put things in chronological order of the recipients pieces of the donor's memories to see if that helps. I'll run through what I have if that's OK?"

"Fine."

"Giuseppe Da Silva's patient who had her pancreas has the memories of Sophie Beckinsale when she attended primary and secondary school. They seem to go up to the point where she sat her A-Level exams. She got Sophie's pancreas. That's the earliest memory we seem to have. I've started there."

"Next comes my patient. He is the younger of the two recipients who got her kidneys. He has memories of Sophie's time at university."

"Your patient is third in order. She received the lungs and remembers Sophie going for job interviews and getting her first job."

"Fourth is Nicola Stephens patient who got the heart. He seems to remember Sophie's romantic assignations up to the point when she got married to Zac Coleman, an American."

"I have placed Isabel Mason's patient next. She got the other kidney and remembers enjoyable holidays in Arizona with her fiancé who became her husband, and his family."

"Sixth is Keith Isaac's patient. He got the liver and remembers all the obnoxious traits of Sophie's husband. His womanising, drinking and violence toward her."

"Seven on my list is Mohamed Khan's patient. He got a cornea and remembers Sophie's husband's death in a car accident. It seems strange that husband and wife both die in road traffic accidents, even if they were five years apart."

"Finally, comes Kasara Doshi's patient. She got the small bowel and seems to have memories of Sophie's life after her husband's death. It appears that her memories stop sometime during the day she died. There are no memories of the actual crash."

"That's the chronology of Sophie Coleman's life based on the memories of eight people who, as far as we are aware, never knew she existed, until they got one of her organs."

"What do you think?"

"Listening to your logic gives me the impression that we have a life story of an individual who had to die to have it recorded by eight strangers. What do you make of it?"

"I can't believe that I'm now going to say what I believe, because it is preposterous, unscientific and downright bizarre."

"I doubt it can be any stranger than what you have already said."

"Wait till you hear it. Do you have a strong drink to hand?"

"I do. Tell me the worst."

"We have a lot of evidence. No, that's not correct. What we have is information, not evidence. There's enough of it to put together the life story of Sophie Coleman, nee Beckinsale. This information comes in the form of what seems to be her memory."

"We know that memory is stored in the brain. The information we have been given by the organ recipients suggests the following scenario."

"On the day of her death, the information we have seems to suggest that Sophie Coleman's memory leaked out of her brain, and into the rest of her body. The leaking memory appears to have become fragmented, with small pieces of it lodged in the organs we harvested and transplanted." Allan heard a gasp on the line.

"As if that was not bizarre enough," continued Allan. "After the transplant, these particles of memory seem to have gravitated from our patients organs back into their brain, to be recalled as we have heard."

"The cycle of her memories seems to be from Sophie's brain to her organs, then to our patients bodies and upwards into their brains. I told you it was preposterous and bizarre."

"You are not joking, Allan. If I did not know you and the history of what we have found, I'd have you committed," said Adil.

"I'm not sure you shouldn't do it anyway," said Allan.

"The problem is, that all eight eminent consultants believe, to a greater or lesser degree, that our patients have had part of the organ donor's memory transplanted with the organ they received. You have described a plausible route for this transfer, but we all know that there is no system, or combination of systems in the human body, whereby this can happen. It's impossible, but the information we have suggests we are wrong."

"We need to speak to the others to work out where we go from here," said Allan in a tired voice of resignation.

"We do. We should arrange a meet as soon as convenient," said Adil.

"I'd like to do it in person, but a video call might be the only realistic way of getting us all together at the same time. We have everyone's email addresses, so I'll contact them. I'll send them a copy of what I have just old you. We need to put our collective heads together. I'll ask them all for dates and make arrangements."

"If we are all to meet in person, perhaps the best place would be Harefield Hospital if Nicola Stephens could organise a room. It would be easier for myself and Doshi to drive there. Those based in London can easily get a train or taxi."

"Yes, good idea. I'll ask what the others think is best." At that, the conversation ended, and Allan Cunningham returned to his daily routine. He would compose an email that evening when he had gathered his thoughts.

23

It's Time We All Met

The other eight consultants had all asked to be kept informed of Allan and Adil's findings. Allan composed an email based on the telephone call he had with Adil.

Colleagues,

Thank you for the time you have given to Adil Patel and me. You have all asked to be kept informed.

On the twelfth of February this year, all eight of us transplanted organs into patients who needed them. The eight organs we received came from the same donor, Sophie Coleman, nee Beckinsale. I have confirmed with the NHSBT in Bristol that these eight organs were the only ones used from this donor.

Around four or five months post-op, all eight of our patients started to express the idea that they had received the memories of the donor with her organ. These memories appear to become more detailed as time went on. It also seems that the recipients only received a small fragment of memory. As far as I can tell, none of our patients have her total memory.

In chronological order, the memories these patients

received are:

Giuseppe's lady who remembers school days.

My patient remembers her university studies.

Adil's patient remembers her first job.

Nicola's patient who remembers her romance and marriage.

Isabel's patient remembers enjoyable holidays in the USA.

Keith's patient remembers her husband's domestic violence towards the donor.

Mohamed's patient remembers her husband's death in a car accident.

Finally, Kasara's patient remembers the donor's life after the death of her husband.

Putting these pieces of memory together means that we can recreate much of the life of a dead organ donor. She might be gone, but she seems not to have been forgotten by our eight patients.

There appears to be only one way I can make some sense of this puzzle. It is this.

On the day of her death, Sophie Coleman's memories left, leached or migrated from her brain into her body. Pieces of her memory were then stored in the eight organs we harvested and transplanted. After the transplant, the memories made their way into our patient's brains for them to recall.

Medically and scientifically, this seems complete rubbish to me. I have no other way to attempt to explain what we are all seeing with our patients.

I am convinced that the only way to make sense of this bizarre situation is to pool our considerable experience. We could meet on a video call or in person if you prefer. Adil suggested Harefield as the most central venue if a personal meeting is preferred. A Video call may be the easiest to arrange.

Let me know what you think.

Allan.

24

A Meeting Of Minds?

A week before Christmas 2013, and ten months after their collective operations, Allan Cunningham was sitting in his office in front of his computer, waiting for seven other consultants to join his video meeting.

One by one, they asked permission to join the meeting. Their faces appeared in little boxes on Allan's screen as he let them in. Soon, they were all assembled on screen. After a round of pleasantries, Allan started the proceedings.

"Thank you all for joining me this evening. The subject of this meeting is without doubt the strangest that I have ever chaired or attended."

"Just to recap. On the twelfth February this year all eight of us performed transplant surgery. The organs all came from the same donor, a woman aged thirty, who had tragically died in a motorway crash on the M25. Her name was Sophie Coleman nee Beckinsale. It has been confirmed by the people at Bristol that only these eight organs were used."

"My own patient received one of her kidneys. The operation was unremarkable, and his recovery initially proceeded as expected. I believe that all of the eight patients are still alive and progressing reasonably well, but not without the odd hiccup. Am I correct?"

There was a chorus of yeses.

"About four or five months after the transplants, all of our patients started to report information about the donor. At first, they reported a vague belief that it was a female donor."

"It is interesting to note that all eight started to recall this information at about the same point after surgery. There are those among us who might consider such a co-ordinated response to be a hoax, perpetrated by eight individuals working in concert. I might be one of those sceptics."

"At this early stage, I believe that we mostly thought that our patients were dreaming or delusional. This is not an unusual occurrence after transplant surgery. However, over time this memory recall became more specific, starting with the donors name. This was variously Coleman or Beckinsale depending on what came next."

"If their supposed memories were of the younger woman, she was called Beckinsale. If these memories were of a time after her marriage, she was called Coleman. The donor was born Sophie Beckinsale and married Zac Coleman, and so it seems to tally."

"Our patients continued to give more detailed information about the donor's life, as time passed."

"I sent you my assessment of the chronology of these memories. There is no known incidence of this ever having happened in the past, anywhere in the world. I have no way of understanding how pieces of memory can move from the brain and become embedded in an organ."

"Even more perplexing is how this embedded memory, if it was transplanted into a new body, can then make the reverse journey into our patient's brains. This circuitous route seems to be the only way that fits the information that we have from our patients, and has no scientific validity."

"Some of the information our patients had might have come from open sources, like the internet. Some of it would not have been openly available. Between them, they seem to have many of the memories of the life of this donor."

"One patient describing the memory of an organ donor would be strange indeed. For eight to have all of the donor's memories between them would suggest a conspiracy. I have no knowledge that they are related or even know each other. For a conspiracy to be considered, they would have to know each other at the very least, even if only through an underground internet group."

"If it is a conspiracy, what is the point? Where is the pay-off and who is the guiding light? We have had nothing in the way of a demand and so, I suggest we can park that theory, unless we do."

"I got my Head of Psychiatry to sit in on one appointment with my patient, as did some of you. I believe that none of those professionals have provided a rational answer. My Head of Psychiatry felt that our patient sincerely believed what he was saying. She could offer no logical or reasoned psychiatric opinion as to his state of mind. I was at a loss as to what to make of the situation."

"A chance meeting between myself and Adil Patel brought this issue to light." Allan pointed to Adil on his screen, not remembering that his audience had a different screen view.

"I was telling him my story when he chimed in with the donor's name before I could tell him who my patient was describing."

"After my initial shock, Adil and I discussed what to do next, which is when we decided to contact NHSBT in Bristol. From there, we contacted all of you and were staggered to find that all eight of our recipients got an organ from this same donor, and we all had the same issue."

"What now? Has anyone any idea how this might have happened?" Isabel Mason had her hand raised. "Isabel. What can you add?"

"Actually, not much except we all know that some organs are associated with emotions by the general public. There are emotions of love associated with the heart. People also get gut feelings or feel liverish. None of that is scientific, and if it is true, then there must be a rational medical explanation, even if none of us know what it is."

"There are also people on the internet who would have you believe that taking their revolutionary drink or medicine cures certain emotional conditions associated with specific organs. None of this has any scientific validity. On the other hand, I believe we all know that some negative emotions can have a detrimental impact on the health of certain organs in the body. This is, as far as I know, a one way system of emotions affecting a specific organ. I have never heard of organs affecting emotions or, as in this case, memory. It is impossible for me to believe that in February of this year, we all transplanted organs that contained a small part of the donor's memory. That sort of thing is science fiction, not medical fact."

Kasara Doshi was next to speak.

"We are all aware that explicit memories are stored in the hippocampus, the neocortex and the amygdala. These are memories that can be consciously recalled, which is what we are talking about tonight. No part of the brain was used in any of our operations. As Isabel says, I can't see how any part of an individual's memory can be housed in an organ other than the brain. And yet, our patients seem to have this donor's memories. Moreover, it appears that our patients did not get all of the donor's memories, just one small part of it. It's as if the donor's memory became fragmented and a small part ended up in each organ that we transplanted. I can't believe that I am saying this, but it's the only explanation that seems to fit the facts that we know them to be."

Giuseppe De Silva was next to interject.

"As Mrs Mason says, there are a number of published papers that seek to show that memory and emotions can be associated with certain organs. They are all suspect in that they use methodology and tests that seek to prove the author's theory rather than seek the true facts."

"Likewise, the internet is full of quacks who sell their own miraculous formula which they claim will enhance or repair emotions or feelings by acting on specific organs within the human body. All of it is rubbish and none of this quackery explains what we all have here."

Nicola Stephens raised her hand and spoke next.

"I struggle to believe what I am about to say, but say it I must. One aspect of what we are discussing is being overlooked. If the donor's memory has somehow been transplanted with the organ, as seems to be the case here, it could a have significant impact of future transplant surgery."

"We could investigate this situation further and we may find some mechanism for this transfer of memory. What then?"

"Before any operation, we speak to our patients about the issues and problems of transplant. We give them the facts. Are we going to tell them that a side effect of an organ transplant is that they may inherit some of the donor's memories, good or bad? I can just imagine that conversation. The patient could refuse the transplant until I can assure them that will only get good memories."

"If this memory transfer is real and widespread, then there are religious and personal dimensions to this issue," said Mohamed Khan, who was next to speak. "Would potential donors dictate that only their religious memories are harvested and passed on? I know it sounds far-fetched, but look at what is happening around the world at the moment. Religion and lifestyle beliefs are rampant in the Middle East, India, Pakistan, much of Africa and the bible belt of the USA. Autocratic dictators would see this as a heaven sent opportunity to control their populations views and beliefs."

"Donors, or their families, might want to perpetuate a particular religious or lifestyle belief into the recipient through the memory of their loved one. What would that do to the availability of organs for our work?"

"Good point," said Allan Cunningham. "I suppose the question for all of us is, do we want to proceed and investigate this extraordinary event, or is it best left alone?"

"We overlook one key element in this story if we leave it alone," said Keith Isaac. "We forget that our patients already have these memories and might broadcast it to their friends and on social media. We can't control that, if we ignore it now. The issue might become public knowledge whether we like it or not. Can you imagine what the media would make of this story? I believe that we need our patients to think that they are the only ones with this problem. It's a psychological issue that only they have. Keep it contained until we identify a creditable medical rationale for it, or are able to write it off as an aberration. We could use our psychiatric colleagues to help us control their desire to broadcast. Make the patients believe they are alone and might be suffering a mental episode. Keep quiet until we find out what it is and treat it."

"That's harsh, but in the circumstances outlined by Nicola and Mohamed, might be the safest option," said Giuseppe.

The formal structure of the meeting soon began to disintegrate as some participants speculated hypothetical scenarios and medical or scientific theories.

The combined IQ of the participants was over one thousand, and their experience spanned many decades of highly complex medicine. Despite their intelligence and training, nothing discussed in this meeting shed any light on the problem. The only consensus from their ramblings was that it needed to be contained within this group. It was too dangerous to be let out. The patients had to be kept in the dark about the others, and those in their teams who knew about the problem needed to stay quiet.

Allan Cunningham brought the meeting back to order. "Ladies and gentlemen, we are going around in circles. Adil, I note that you have kept your powder dry and said little so far. You look like you might have something to add."

"I think we may be approaching the problem from the wrong angle. I don't think anyone at this meeting tonight would describe the situation with our patients as anything other than unique." There was much nodding of heads on computer screens.

"We eight have many years of experience of transplants, but no experience of anything like this. It seems to me that our combined medical and scientific knowledge is totally useless. Listening to what has passed so far, shows that we are all in the dark."

"Perhaps this situation is unique because there is something quite unique about our donor. Maybe the answer to this conundrum lies with the life of our donor."

"It seems that our patients have all received a different part of the donor's memory, along with her organs. From listening to you all, it also appears that none of them knows everything about this lady. We could get them all together to relate what they know. If we do, they will know that they are not alone. That could create a very large problem for us as transplant surgeons. Still, my suggestion is that we somehow put together as much of this donors life as possible. That seems to be the most logical sensible way forward. Perhaps there is something in the donor's life that will give us a clue as to why this has happened, and why she is different to every donor we have ever encountered. Maybe our situation is unique because this particular donor just happened to be unique in some way that we do not yet understand."

"Thanks, Adil," said Allan. "That is a very valid observation. None of us can think of any neurological, medical or scientific way in which this could have happened. Perhaps the only way to unravel this mystery is through the life of the donor. I can think of no other possible approach. Does anyone have a dissenting view?"

"Adil is correct," said Nicola Stephens. "Perhaps the answer lies in the donor's life. Given what we have discussed this evening, it is the least scientific but most sensible suggestion I've heard so far. I think that it is the only way forward."

"I agree with what has been said so far about getting these patients together in the same place. It would be a logistical nightmare and could unravel quickly," said Allan. "I suggest we all devote our next appointment with our patient to this question of their transplanted memory of the donor. Use our psychiatric colleagues to get them to reveal as much information as they can muster. When it started, why do they think it's real, is it associated with anything in their own lives?"

"One patient having hallucinations about the life of their donor I can just about cope with. Eight patients, connected only by the same organ donor having memories of that donor is beyond my comprehension."

"We should record their next appointment and put them all together to see if there is a possible answer. Do you all agree?"

Allan asked each participant if they agreed. They all did, as none had any better suggestions.

"I would suggest that you send your recording, or a transcript of it to me. I'll get my research registrar, Graham Bradbury, to pull it all together to see how much of this donor's life we can recreate. Are you happy with that?" They were.

"I'll contact you all again in the New Year when Graham has completed his work. He will be delighted to be involved, as he was the first member of my team to highlight my patients issue."

The meeting concluded with many Merry Christmas and wishes for a Happy New Year exchanged.

25

Interview Protocol

After the meeting, Allan exchanged several emails with his eight consultant colleagues.

If they were going to try to get to the bottom of this mystery by asking their patients to recall all of the memories they had from Sophie, they needed to do it in a controlled, consistent and scientific way. They all needed to approach their interviews in the same way to get as consistent a response as possible from their patients. They required a strict protocol for the patient interviews.

Allan and Graham set about producing a draft protocol for the others to inspect, amend or improve in any way appropriate. It did not take many draft proposals before the eight eventually agreed on a protocol. They then set up their appointments as agreed.

The first step in the protocol was that each consultant would contact their patient by telephone. They were to explain to the patient that this transplanted memory was a situation that was unique to them and needed to be investigated further.

They could point to a possible unknown psychological issue with either the patient or donor if anyone asked. They would tell their patients that there would be a specialist in the interview who was qualified to help if needed.

The patients would be told they would undergo the usual sequence of medical tests to look for clinical anomalies in the first instance. After these routine tests at this appointment, the final stage would be to conduct a taped interview. During this interview, the patient would be encouraged to give as much information and detail as possible about their memory experiences. When did it start? Why did they believe it was the memory of another person and not their own? What part of that memory do they think they have, and how much detail do they have?

The point of this conversation was to make each patient think that they alone had this problem. They would be encouraged to consider that this memory they claimed might have a psychological basis. They would be encouraged to believe that it was in their interest to keep it to themselves until a solution was found. If it was a psychological issue, their medical team needed to know so that they could explore it further with appropriate specialist help.

For that reason, the appointment would be longer than usual. Each patient needed to consent to this approach.

All eight individuals agreed when contacted, but it was evident that some of them were experiencing anxiety about their situation.

After their usual tests to check the health of the organ and their patient's physical well-being, the patients were to be escorted to a quiet side room where the interview would be conducted.

The tape was to be switched on, and the date and time were recorded. The occupants of the room identified themselves. Medical personnel added their specialities and qualifications. It was almost like a police taped interview. This was a deliberate ploy. It was included to make the patient feel this was a formal and systematic process being followed.

The patients were then briefed on what was expected of them and that the information might be shared with other qualified personnel. People who had an interest in this field of psychology. The patients were informed that the reason for sharing this information was to have as many fresh eyes on the issue as possible to try to understand what might have occurred. They were not to be told that other patients had the same experiences. The patient had to agree to the information being shared. If any were to disagree, the interview would be terminated.

Once their agreement was recorded, the patients would be asked to start recalling their donor's memories, starting as early in their life as they could remember. They would also asked to give as much detail as they could recall. Nothing was to be considered too trivial to be included.

At the end of the interview, the patients would be asked to confirm the extent of their memory. The tape would be switched off after recording the time it ended.

The eight consultants also agreed to send as much personal and medical information as they had about their patients. They could not know whether these issues arose in the donor or their patients. They needed to examine every possibility.

The interviews and patient reports would then be sent to Graham Bradbury at St George's Hospital.

Graham was asked to request Sophie's notes from Chase Farm Hospital and the reports from the two organ harvesting teams. He expected to have a significant quantity of paperwork or extensive information in the inbox of his email account. This might become a large operation. He received permission from Allan to ask for help from members of his surgical team. He would only ask those already aware of the problem of transplanted memory to volunteer. No point in adding to the number of people who already know of this issue.

26

A Life In Eight Memories

As each consultant had agreed to re-examine the files and lifestyles of their patients, they all produced reports that were as comprehensive as they could muster. Graham received them throughout January 2014.

A great deal of medical information had been sent by email. It needed to be printed to enable everyone involved access.

Graham laid them out in order of patient age rather than memory age.

Adil Patel's patient, Abigail Gould, was the oldest to receive an organ. Adil reported that Abigail was born Watson in Norwich in 1964. There was nothing of note happening in Norwich at the time of her birth or during her school years. By the end of the 60s, men had landed on the moon. That could not have affected Abigail or her ability to absorb another's memory.

She had one female sibling with whom she got on well.

She started to smoke at secondary school and continued until she was told and shown that it was killing her. Her situation was worse than most smokers because she worked in a flour mill in Bury St Edmunds. There, she was also exposed to the dust of the milling process. When this was combined with her smoking, it accelerated her medical descent to the point she needed a transplant.

In her younger days, she had taken a two-week package holiday to Spain in 1989 for her honeymoon. It was the only time she had ever been out of the UK. There had been some holidays in England, Scotland and Wales.

She had taken the contraceptive pill, as did many women her age and most now did.

Her schooling and general lifestyle were unremarkable. The fact that she became a heavy smoker was not unusual for young people at this time.

She had no particular religious fervour and attended church only at Christmas for the midnight mass. She was not a witch, nor did she participate in seances or any other new-age mumbo jumbo. She had not smoked pot or marijuana or taken any other mind-expanding drugs in her youth or teens. Her IT competence was below average for people of her age, Adil reported.

Adil could find nothing in her personal history that would give him any clue as to why she was different to every other organ recipient. As far as Adil could tell, there was nothing in her husband's lifestyle to report.

Adil confirmed that he and his team had examined Abigail's medical history, biopsy and test results. There was nothing there that they did not expect to see. Other tests might throw up an abnormal reading, but knowing what more to test was a mystery.

Adil's conclusion was that there was nothing about his patient that would give any clue as to why she had the donor's memories.

Keith Isaac's reported on the next eldest, Jack Gibson. His medical history was more extensive but equally vague.

He was born in a poor area of London, where he lived with his parents and three siblings. His family were also poverty-stricken. As a child in that part of London, Jack was subjected to familial violence from his father and occasional violence from police and other so-called authority figures. A clip around the ear to wayward children was routine punishment from the police, school teachers or shopkeepers.

Jack had become involved in petty criminal activity at an early age. He was not proud that he had. It was a way of life for children like him from poor homes.

He joined the army to escape the Metropolitan Police, who were trying to establish that he had been involved in a robbery that had turned violent. Jack denied that he had taken part in the violence, but he confirmed to Keith that he was guilty by association. He was a member of the gang that committed the crime. Keith was sceptical of his patient's claim of innocence but had no evidence to back up his suspicions.

In the army, Jack Gibson saw combat in Iraq and Northern Ireland. He had a diagnosis of PTSD some considerable time after he left the army. His diagnosis was only confirmed after he had suffered the consequences of his excessive drinking. His wife left him because of his aggressive and occasional violence towards her and their children.

During his service in the army, Jack Gibson had several encounters with authority that stemmed from his aggressive behaviour. Over his service, he spent several weeks in what he described as the glasshouse - army prison.

Keith pointed out that before his being put on the transplant register, Jack had demanded to know details about any organ donor. Keith indicated that he thought this might have been a racist request, which he did not pursue. He told his patient that he would only get the same information about the donor as his team. He would be told if there was any medical issue that might impinge on the transplant. He warned his patient that if he demanded to know any further details, he would be removed from the register.

Perhaps military combat, parental violence, PTSD and racism might be areas to explore for this patient, but that was not somewhere Keith was qualified to do or where he wanted to go. That was best left to an expert in those fields. Keith worried that might mean adding more people who knew about the issue.

Aside from these caveats, Keith reported that his team's investigation of Jack's medical files indicated no post-operative abnormality that might shed any light on the problem. He made it plain that his team did not find Jack Gibson to be a particularly likeable man.

Keith had added nothing substantive to the resolution of the problem.

Next came Isabel Mason's assessment of Rose Prentice's life and medical files.

Born into a middle-class family in Essex, Rose's life was a product of the times.

She lived through the Thatcher years and had female friends who worked at the Ford Motor Company factory in Dagenham. Some of Rose's friends had been involved in the women's employee's strike for equal pay. There was no way this could influence anything other than Rose's sympathy for her friends and a hope for a better deal for all women.

The only holiday she talked about was a trip to Southend on Sea with her parents. She had once travelled to Spain on a package holiday before her condition precluded getting medical insurance. She liked the sun but not the food.

She had never been to the USA and only started to worry when she had memories of somewhere called Sky Harbour. She had no interest in sailing. She could not understand why a harbour would be associated with the sky and flying.

She only learned that Sky Harbour was the name of Phoenix Airport in Arizona when her husband looked it up on the internet.

Why she was able to describe a street and house in a town called Sedona a hundred miles north of Phoenix was a mystery to her. It seemed a lovely place. It was somewhere she would have liked to have visited but never had. She even had a memory of a visit to the Grand Canyon, which is within easy reach of Sedona by car. Apart from the magnificent sight, Rose remembered how quiet the place was. She thought this strange.

Isabel reported that Rose had had two treatments of shock wave therapy to break the kidney stones down to allow an easier path to flush them away. This was Rose's only significant departure from pharmaceutical treatment. Isabel could not believe this therapy had any bearing on possible memory transfer. None of the other patients had this treatment, so how could it be implicated?

Isabel Mason's report on Rose Prentice mirrored the report Graham had produced about Richard Ainsworth.

Rose had a few extra issues consistent with her age, but nothing noteworthy. Both managed their conditions as well as could be expected. Graham knew that Richard had multiple absences from school but that his education was largely unaffected.

Isabel confirmed that Rose's education had suffered as a result of her medical condition and her frequent hospital visits and treatments.

Isabel did not find anything in Rose's medical records or personal life that would throw any light on the memory transplant.

Kasara Doshi's report on Megan Thomas's life and medical history was as unrevealing as the information from the other patients.

Megan was born in Nottingham in 1970. Her family were originally from Wales but had moved to Nottingham before Megan's birth. The family lived in the Hyson Green area of the city. Wales is widely known for mysticism. People like Arthur Pendragon (King Arthur), the druids and Merlin the Magician are all magical Welsh legends. There is no hint that Megan's family was descended from these people or had any supernatural powers.

She worked for Boots, the giant pharmaceutical company, as did her husband. She was a merchandiser and had no contact with drug research except through her husband, a chemist also working for Boots.

Her husband worked in the research department of the company in Nottingham. No one knew if he was researching new compounds. He might have carried particles of these compounds home when he left work. It was a remote possibility and probably of no consequence.

Megan's only medical unknown issue was a gastric upset she got while on holiday in Egypt. There is no record of the treatment the local Egyptian doctor gave her. Karsa's medical opinion was that Megan had all the classic signs and symptoms of amoebic dysentery that were prevalent in North Africa. He did not know what treatment the Egyptian doctor gave Rose as she did not ask.

He had done a CT scan as part of his initial investigation, and a re-examination of that scan had revealed nothing more than he already knew.

Megan had had three operations on her bowel before her transplant. The scar tissue was an irritant during the transplant but not a significant problem.

As far as Kasara could ascertain, Megan Thomas's lifestyle and medical history were as expected.

The report from Professor Mohamed Khan on Arthur Morgan was the shortest of the eight that Graham would receive.

His birth and childhood in Wales were as irrelevant as it was ordinary. It would be possible to ascribe Arthur's ordinary life in Wales to any child in that country or anywhere in the rest of the UK. As one might expect of a Welshman, Arthur played rugby. He described himself as a run-of-the-mill club player.

University education at Bath was followed by working as a social worker in London. He had nothing in his history to suggest anything that might not be expected. As far as Mohamed knew, Arthur had not contacted any transplant patients in his work.

Mohamed had no clue as to why the transplant of a cornea would allow his patient to see any part of the donor's memory of life.

Nicola Stephens reported that the cause of Jacob Marshall's heart condition was as equally unknown as his ability to remember part of Sophie Beckinsale's life and marriage.

Jokingly, she suggested that supporting the Watford Football team might have harmed his heart, considering their inconsistent form in the league. It may be a joke, she wrote, but it's no less ridiculous than him having received part of the organ donor's memory.

She saw nothing of note in his life and upbringing that is worthy of mention. His immediate post-operative recovery was delayed a day by the discovery of a tiny bleed. The rest of his recovery, with its ups and downs, was not unusual in any patient who had undergone a heart transplant. She reviewed all of his test results before and after his surgery. None of the tests or biopsies showed anything unusual.

There is nothing extraordinary in any of his pre or post-operative tests. I am sorry that this young man's history gives no clue how this problem might be solved.

Giuseppe Da Silva reported that his patient, Simone Jessop, was born with type 1 diabetes that ran in her family. She was a careful recorder of details about her condition. Despite her illness, Simone had done well at school. She had a friend in her school who was also a type 1 diabetic. The two were very supportive of each other.

Simone's family life was blighted by the fact that her mother had died as a result of type 1 diabetes. She was fourteen when this occurred. Her family life after this point was quite hard, but Simone and her father coped very well.

The only other thing of note in Simone's life was that she was very computer-literate. She was like most other teenagers in this regard. She also kept meticulous computer records of her diabetes. If there was a computer-led conspiracy among the patients, Simone might be the conduit. She did not give the impression of being inclined that way.

Reading through the seven other reports, Graham could see nothing in any of them that might give him clues on how this had happened. He noted the mention of Simone Jessop's computer literacy. That might be an avenue to explore if a hoax or scam was considered.

All of the medical tests were what might be required for the patient's conditions, and none appeared to be unusual in the circumstances.

None of their social backgrounds stood out except that of Jack Gibson. He had faced combat and had PTSD as a result. His domestic violence was not uncommon in the circumstances. One out of eight with a similar social background to the donor was not noteworthy.

27

Where To Start?

Before receiving the tapes and transcripts of all organ recipients, Graham had started to work out how he would approach the problem when they arrived.

The science of life is called physiology. It's an experimental science that seeks to help us understand the mechanism of living things such as the human body. It aims to look at everything from the molecular and cellular level up to the whole body's function. It was unfortunate for Graham that it is still an experimental science since he considered this speciality might offer him some help. Given the complex nature of the problem that he and the eight consultants faced, Graham thought this was the best approach.

Somehow, eight recipients of organs all had a similar experience. They all seemed to have received a portion of the donor's memory with one of her organs. None of the memories seemed to overlap, and no one had the combined memory except for Graham, who had it on paper.

There was no medical evidence as to how this could

have happened or that it had ever happened previously. Graham was in totally uncharted water.

He knew that memory encompasses the facts and experiences that people consciously recall to mind. It also has ingrained knowledge that surfaces without effort or even awareness, such as the ability to breathe or talk. It is a short-term pool of information and a more permanent record of what we have learned.

The real problem of memory is not that we lose some of it as we age, but scientists are still not entirely sure how it forms and how it is stored or recalled. Scientists know that memories are created, stored and recalled, but more than that, the picture becomes hazy.

Doctors know that a range of neurotransmitters in the brain play a significant role. A nerve does not reach from the brain to the foot like a long piece of string. The pathway is filled with billions of neurons. None of which is in direct contact with its neighbour. At the end of each neuron is a mechanism known as a synapse. This is like a giant telephone exchange. Synapses shoot neurotransmitter chemicals across the gap between the neurons to the next in the chain. The nervous system acts like a series of relay stations passing messages to where they are needed. Current theory says that this movement of messages along the relay strengthens the synapse and consequently helps build up memory.

There is also evidence that recalling a memory from the past means that when it is returned to the store, it is altered by our current perceptions. As a result, a restored memory may not be an exact copy of the one that was made at first. Memories reflect real-world experience but with varying levels of fidelity to that original experience. Were the memories experienced by the eight transplant patients, the originals or near copies, altered by time and recall? Who knew? Certainly not Graham

A couple of nurses on his team had spoken to Graham about a few transplant patients undergoing personality changes after their operation. It did not happen to all patients, only a few. Could these changes be attributed to the new organ? No one knew, but it was worth mentioning.

Exactly a year after Sophie Colman's death, Graham was mulling over his problem and how might be the best way to approach it. It suddenly came to him that if he could identify a mechanism for what had happened to these eight patients, he might be in line for a Nobel Prize. The sudden euphoria was equally quickly dispelled. Even to be nominated for the prize, Graham understood that he would need to uncover the process. At this moment, he was uncertain where to begin.

"I'm thinking about the accolade I might get at the conclusion of the process, when I can't even work out where to start." He muttered to himself.

Graham started by thinking about what was known about the donor and what was not. He decided to start with her.

He initially dismissed the information coming from the recipients. There was no known mechanism for memory to migrate from the brain to any other random part of the human body. There is order in that process, and the recipient's memories from random organs are certainly not part of that orderly system. Until proven otherwise, the recipient's stories had to be discounted. The more he thought about it, the more stupid he felt. Everything that he was now doing was based on these stories. He could not dismiss them.

"You're examining the validity of these stories and the first thing you do, Graham, is to discount them," he muttered. "That's just silly and unprofessional. Much as you don't want to, Graham, you need to use them." Graham did not, as a rule, talk to himself, but this puzzle had him at it, big time.

Graham had received Sophie's GP medical records. They showed nothing remarkable except for the hospital report of the beating she took on the night her husband died in a car crash. On paper, it looked like a frenzied attack. There were lots of contusions and a few cuts, but no broken bones.

Graham turned his attention to the treatment she received in Chase Farm Hospital A&E. She had arrived unconscious and largely unresponsive but still alive. She had clear cerebrospinal fluid leaking from the ear and nose. A sign of a Basal Skull fracture.

X-rays showed the fracture, and a CT scan showed that she also had three countercoup brain injuries. All three injuries were in parts of her neocortex where general memories are stored.

Countercoup injuries occur when the soft brain is smashed against the hard skull, as might happen in a car crash. That might be an area to explore, thought Graham.

Could her head being smashed around in her skull have dislodged small parts of her neocortex containing her memories? Graham could not see that if this did happen, how did these bits of neocortex get into the eight organs that were transplanted. The only possible way for them to get there would be to travel along her nervous system or blood vessels.

Why would they not continue around the body along her nerves, veins or arteries? Why stop at an organ? Had those pieces of memory got to the organ when she died and stayed there when she was brain dead, or her blood stopped flowing?

The more he pondered the issue, the more he thought it impossible. When Sophie was declared dead, her body was kept artificially alive so that the harvesting team had fresh meat. That meant her blood was kept circulating until the harvesters opened her up. However, her brain died before her circulation stopped. Chase Farm would have to confirm brain death before the harvesters could be called to begin their gruesome work. There is a gap between brain death and the cessation of blood flow of about an hour.

Brain death is the best criterion for organ donation. This is usually a catastrophic or irreversible injury. A brain-dead individual shows no clinical evidence of brain function upon physical examination. So, no movement of charged ions around the body is possible. The only movement around the body after brain death, and up to organ harvest, is mechanically generated blood flow.

Sophie's memories were circulating around her body

via her nervous system before stopping in eight organs when her brain activity ended. Or, Graham wondered, did pieces of neocortex keep circulating until her blood flow was stopped during the organ harvesting procedure. Was there something in the harvester's report that might shed some light? What if Sophie's memories were coursing throughout her body as neurotransmitter chemicals in her nervous system and particles in her bloodstream? Her memory circulates in both systems. The mere idea blew Graham's mind.

"Think Graham," he muttered to himself. He re-examined the notes he had taken.

Sophie Coleman was pronounced brain dead at one forty-four in the morning. For that to have happened, her neural function would have stopped a short while before. The organ harvesters took just under an hour to get there and start their work. At that point, her blood circulation would be stopped. So, between half past one and roughly half past two in the morning of 12 February 2013, both possible pathways for memory leakage had closed down.

Memories do not float around a human body like unstable free radicals looking for a home, thought Graham. Memories emanating from the brain give us humans the skills of how to walk, talk and do other sentient tasks. They do not wander aimlessly about the human body and lodge themselves in a pancreas, just for the hell of it or because it seems like a nice place to stay.

When he read the hospital A&E and harvester's accounts, Graham saw nothing he did not expect to see. Up to this point, Graham had seen nothing in any medical reports that might give him a clue or a place to start.

Another factor to be considered might be the nine hospitals and the medical teams involved in the accident and transplantations. Graham was a member of one of the teams that performed a transplant. He didn't see anything unusual in the procedure, considering its complexity.

Graham was a very experienced surgeon and an expert in kidney organ transplantation. He had to admit to himself that his knowledge of the processes and procedures in the other transplant operations was not his forte.

Still, all eight consultants had over one hundred years of experience between them and were unlikely to all make any mistake on the same day with the same donor organs. It was just not conceivable.

The only places left for Graham to consider were the eight recipients, their operations, and their lifestyles.

28

Fact Or Speculation?

By late February or early March 2014, Graham had received the eight written transcripts of interviews and the tapes that went with them. He also had Sophie's general medical file, her Chase Farm A&E file and the report of the harvesting teams that removed her organs. He had all of the information that he needed, he thought.

The consultant appointments were focused on gathering as much information as possible that the patients had about Sophie Beckinsale or Coleman. This was done following the agreed protocol.

He had already done some work on the medical files relating to Sophie and her death. All that he had left were the tapes of the interviews with the eight recipients of her organs, including the one he made with Richard Ainsworth.

Graham's plan was simplicity itself. Begin at the youngest age that any recipient remembered Sophie's life, and work through the rest of it as best he could. He established fundamental markers such as childhood, school, university, first job, marriage and death. Before considering any of the transcripts, he proceeded with his plan. He already had an idea of where they all fit into the scheme.

Graham laid all the transcripts on his office table in sequence, starting with the earliest memory of the donor's life. If the answer to this problem lay in the life of this donor, his job was to make as much sense of that life as possible. He needed to piece together a biography of Sophie Coleman to see if he could identify any possible connection with her life and the memories she seemed to have passed on.

None of the transcripts were equal in length. The one from Simone Jessop, Giuseppe Da Silva's patient, was the longest. Simone was also the youngest patient in the group and had memories of Sophie's school days. She had not long left school herself. He wondered if that was relevant to the memory that she had acquired.

Graham started to wonder if the memories of events in Sophie's life that other recipients acquired were related in some way to their own life experiences. He also questioned whether the length of the transcripts was related to the patient's mental or memory capacity.

He thought he would almost certainly have to consider the eight patients' life experiences, age and mental capacity. As he considered this, he was not convinced there was enough information in the files or transcripts to give him what he might need. If it was decided imperative by all of the consultants, then a further set of patient interviews would probably be required. The longer this exercise took, the more detailed the interviews they had with the patients, and the more chance this information might escape into the public domain. Keith Isaac was correct. This could damage future organ transplantation for a very long time.

When he played the tapes, he was interested to hear the voices of the other patients as they spoke. He thought it might give him a better sense of who they were. It made his efforts more personal than reading a paper transcript.

Beginning with the memories of Simone Jessop, who remembered Sophie's primary and secondary school days. Graham knew she was only eighteen, but she sounded mature for her years. He wondered what she and the others looked like.

Graham listened to the tape and made some initial notes on what might be in the public domain and what almost certainly was not. Simone had mentioned her primary and secondary schools by name. Sophie's attendance at these schools would be confirmed by a telephone call.

Simone also mentioned school friends called Amy Bright and Joyce Fisher. She also indicated that Sophie particularly liked a boy named Isaac Bell, who was a bit of a bully to Joyce. Identifying these three might be more problematic. They were added to his long To-Do List. This was an area where he felt he might need the considerable help of the conscripted team volunteers.

He moved on to the memories of Richard Ainsworth, who had information about Sophie's life at the De Montford University in Leicester. As with the previous transcript, some would be in the public domain and some not.

It was during her time at De Montford University that Sophie learned of her parent's deaths. They had been hit by a youth in a stolen speeding car as they walked through the town of Smisby. They were both thrown a considerable distance and killed instantly. Sophie was now an orphan, all be it aged twenty.

Her course, student accommodation and regular eating places would be easily identified. A lady called Neo Masisi from Botswana and Joan Tapp, who was in the next room and studied Chemistry, might be a bit harder. More fodder for his To-Do List.

Third to be examined was the testimony of Abigail Gould. She seemed to have information about Sophie's first job after leaving university. She talked about Sophie having two interviews. The first was a preliminary interview to get her onto the next stage. The second stage was a full presentation before she got the job.

Abigail had mentioned the name of the company and the names of some of those on the two interview boards. They might be easily checked if the hotel company would play ball.

Fourth was the testimony of Jacob Marshall, who had information about how Sophie first met and married her husband-to-be, Zac Coleman. He reported that Sophie nearly went to bed with Zac on the first night they met. She was captivated by his eyes and dazzling smile. According to Jacob, the romance was what might be described as turbulent. It was very physical, he claimed. Sophie and Zac were married in the church at Smisby only because she still had some friends there. Zac got to pick the venue for their honeymoon, Marrakech. Zac's parents were the only relatives of Zac's that attended the wedding. They did not seem pleased with Zac's choice of bride.

Rose Prentice provided information about Sophie's in-laws who lived in Sedona, Arizona. Zac's parents were very right-wing Republicans. Shooting Mexican immigrants as they crossed the Rio Grande was a satisfactory anti-immigration policy. They were not keen that Zac had fallen in with a colonial foreigner. They had fought wars to rid themselves of the British.

Rose could recall Zac's parent's hostility towards Sophie when she stayed in their home in Sedona.

After that came Jack Gibson, who seemed to have information about Zac's extramarital activities, domestic violence, alcohol and drug use. He remembered Sophie finding clothing and jewellery in their bedroom that didn't belong to her.

Jack also remembers Sophie being told that Zac and two friends had killed a Mexican immigrant and got away with it. He threatened Sophie with some of the same if she ever went to the police or hospital and accused him of beating her. Jack claims that Sophie was convinced that he would carry out his threat. She spent her married life in fear for her life.

Then came Arthur Morgan, who had information about a lady named Charlotte James.

Charlotte knocked on Sophie's door one evening and

introduced herself. She said she had come to apologise and warn Sophie about her husband.

She told Sophie that she had met Zac at a trade fair. She was smitten by him and accepted his offer to go to his room. In his room, Zac became aggressive and violent towards her, and Charlotte wanted to leave. Zac refused and basically raped Charlotte. Charlotte did not consult the police as she had been seen going to Zac's room of her own volition. It would be she said, he said story.

Sophie was angry at first but soon became supportive of Charlotte.

Sophie confirmed that she knew of Zac's frequent infidelities and violence. She had found items of clothing and jewellery in her house that were not hers. Sophie showed Charlotte her own bruises from a previous beating and recounted a story of Zac and some friends beating a Mexican immigrant woman to death in Los Angeles.

Charlotte indicated that she would be a witness if Sophie was to bring proceedings against her husband.

Charlotte and Sophie became friendly as a result of their shared fear of Zac and Charlotte's support and sympathy for Sophie's plight.

Arthur's memories of Sophie's life come to a halt after the conclusion of the police investigation into Zac's road traffic accident.

Charlotte James was staying overnight with Sophie when she learned of her husband's accident.

Arthur knew that the police had seen Sophie's bruises after Zac had beaten her the previous evening. They had sent her to the hospital to be checked, and a report was filed with the police.

Arthur's final recollection was of the police coming to Sophie's home to inform her of the findings of their inquiries. It was ruled an accidental death while under the influence of alcohol and cocaine, exacerbated by the driver not wearing a seat belt.

Finally, Megan Thomas had information about Sophie's life after Zac's death. There appeared to be some guilt and anger for some time afterwards. Megan could not get a clear handle on why Sophie had these feelings. They seemed to simmer under the surface.

Megan remembered a few dates Sophie had with men. They did not happen for nearly two years after Zac's death.

Megan's memories seemed to stop at a point where she was arguing with someone called Roberto Caruso in a hotel somewhere in Suffolk. There were no notes anywhere regarding the crash that led to Sophie's eventual death.

In his endeavours, Graham had the added help of one of the surgical teams' junior doctors and two members of Allan Cunningham's transplant patient support team. Everyone was interested in trying to solve this perplexing mystery. They all knew the issue. They had been sworn to secrecy after the reason for it was explained. They were excited to be asked to help, even if it meant giving up some of their free time.

It took several weeks to collate the information into a timeline of Sophie's life. There were several more weeks of work trying to verify details. Some were easy to obtain, and some were not. Those that were not had to be taken at face value because there was no way to verify them.

Once Graham was happy that he and his team had all the information in the correct order, they produced Sophie's life story, or at least as much of it as he could.

He set the information down in a document that resembled a scientific study paper.

Graham and Allan decided that their investigation was to try to find some chemical, physical or emotional rationale for what had happened to eight transplant patients. As scientific men, they completely dismissed the notion of this being a result of anything outside the realm of science.

Graham and the team carefully noted the source of the information they had verified. If their findings were ever to be reviewed, readers would have to know the source of their verified information.

As an appendix, he indicated those pieces of information that had been described but for which he could find no verifiable source.

He pointed out that no patient recalled any memory of Sophie before she started school. Despite this, he concluded that, as far as could be ascertained, Sophie Beckinsale led a normal childhood right up to a time some months after her marriage to Zac Coleman.

At that point, Zac's drinking and recreational drug taking started to have a negative and violent impact on her life and emotional well-being.

She spent much of her married life afraid and

ashamed that she had married a drunken, womanising, violent man.

Having studied the transcript describing Sophie Coleman's life several times, Allan and Graham concluded that the only part that stood out was Zac's behaviour towards Sophie. Everything before and after seemed as normal as any other organ donor they had ever known.

The patient's medical and personal records did not clarify the problem. Graham produced a biography of Sophie's life in case others might read it and discover something he did not see.

29

Sophie's Biography

Graham's recreation of Sophie's life was limited by the memories of the eight transplant patients and what he could glean from the internet. The report he wrote was the sum total of all he could find. In the report, he admitted that there were gaps in her life and that the answer to this problem might be in the gaps. He had no way of knowing.

On April 23, 1983, Paula Beckinsale gave birth to a daughter she and her husband, Robert, called Sophie. As Sophie was born on St George's Day, the couple toyed with calling their daughter Georgia.
Robert Beckinsale ran a Real Estate agency in Ashby, and his wife had been an administrator in the business before she had Sophie. They were what would be termed the middle class.

The family lived in a small village called Smisby, on the northern outskirts of Ashby-de-la-Zouch in Derbyshire. Official records show that Smisby was recorded in the Doomsday Book. The hamlet belonged to Nigel of Stafford and had only five villagers. In 1983, fewer than two hundred people lived in the village and outlying farms.

Its main claim to fame is the nearby area called Tournament Field, mentioned by Sir Walter Scott in chapter seven of his novel Ivanhoe.

This information was easily obtained in the official town records, Graham wrote.

Simone Jessop, who had Sophie's pancreas, has no memories of Sophie's childhood before she started to attend school. She had no memories before this age, and neither did any other recipient. Her memories of Sophie's toys and books came from what was in her bedroom when she started school. Simone seems to have a clear picture of how the room looked. Her recording almost sounded as if she had sat in the room.

Sophie, the child, had three Barbie dolls, two Cabbage Patch dolls, a Tom Kitten and a teddy bear called Benjamin. These were all toys that females coveted at that time. It appears that Sophie also had a fascination for the books of Beatrix Potter and had quite a collection of her books in her bedroom. Simone's memories contain the titles of some of Potter's books.

The bedroom itself had a great deal of pink in it. Sophie had a white throw over the bed, an alarm clock and a small bedside lamp with a mainly pink shade. She had a cream rug on the wooden floor.

Sophie's parents took her to the Lake District to visit the Potter house and estate during the Easter Holidays when she was eight years old. In time, Sophie's obsession dimmed as she found other avenues of enjoyment, just like her contemporaries.

Her primary education was done at a local Church of England Primary School. Her best school friends were Amy Bright and Joyce Fisher. She also liked a boy called Isaac Bell. She thought him very handsome. Amy also liked him, but Joyce did not. Isaac was a bit of a bully and picked on all three girls, especially Joyce.

The three named children might have been at the school, but no records from that period survive.

The girl's favourite music track was Breathe by Faith Hill. They also had a liking for country music.

Sophie concluded her childhood schooling at the Ashby Comprehensive School, which had a sixth form where she studied for her A-level exams. She left with Three A levels, including Maths and English, plus six GCSEs. This was sufficient to get her entry to her University of choice, De Montford, in Leicester.

At this point, Sophie's story is taken up by Richard Ainsworth, who had one of Sophie's kidneys. He is two years older than Simone Jessop. He remembers Sophie's time at University, where she got a 2:1 (BA Hons) in Business and Marketing. This was easy to confirm through the University.

She was housed in student accommodation in the Glassworks, where she had one of the six bedrooms in her flat. She had one good friend and one overseas acquaintance at University. Her friend was Joan Tapp, who lived in the next room to Sophie. The University verified that Joan Tapp was there at the same time and that she lived in the same building as Sophie.

Richard also indicated that Sophie knew a Neo Masisi from Botswana who was studying for an MBA.

We have confirmed that this woman also studied at De Montford University. She was there during the same period as Sophie and has since returned to Botswana. This lady is no longer resident in the UK. Richard must have her details from Sophie's memory.

Richard also informed us that Sophie liked to eat at the Kimberly Library Cafe just off Gateway Street and the Riverside Cafe and Food Village off Newarke Close in the Vijay Patel Building. She apparently went to pop-up events that were held there. The venues and events can be verified by internet searches. Sophie's attendance is possible as they are popular eating venues for students. This was reported by the University.

Sophie liked to eat out as she was not a very good cook. She had cooking facilities in her room that she could use. It was mostly for reheating cheap ready-meals from the local supermarket. Richard's memory of these events is taken at face value as we have no supportable information.

The University confirmed that they had occasional tickets for Leicester Tigers rugby matches. Richard tells us that Sophie attended some of these matches. She was particularly enamoured of their blonde flanker, Lewis Moody and full back, Tim Stimpson. Richard adds that Sophie liked how Martin Johnson, Leicester and England captain, bossed people about. They were all big, pugnacious men. That might help identify why she eventually married the American Zac Coleman, who, it seemed, was built in a similar mould and had similar aggressive traits.

The men Sophie admired in the Leicester team are well known and can be found on any search of the club's website.

The two youngest recipients of Sophie's organs have memories of the first two parts of her life. Abigail Gould, at fifty years old, is the eldest recipient of an organ. She takes up the story of Sophie's life. It seems clear that Sophie's memories are not in concert with the ascending ages of the recipients of her organs.

Abigail remembers Sophie getting her first job in the operations function of a large chain of American-owned hotels based in west London.

Abigail says that Sophie went for two interviews at the headquarters offices of the chain near Denham, Buckinghamshire. The first interview lasted almost an hour. She was questioned by a panel of four senior managers. The second was with a presentation to a smaller group. Abigail mentioned several members of the interview boards by name. The company has confirmed that these people did work for the company at the time. They might well have been on the panels that interviewed Sophie.

Sophie started in the company as a junior business and marketing assistant to a regional director. This necessitated her travelling to all of the hotels in the southeast region of the UK. She spent many nights away from the headquarters office and home. She also attended seminars and trade fairs to represent her company. She slowly worked her way up the corporate ladder.

Sophie had a flat in the Croxley Green area on the outskirts of Watford.

Abigail also stated that Sophie had a few short-term

affairs with some men, one of whom was married and employed by the hotel group.

Abigail seems not to have much more in the way of information. I can confirm that Sophie did work for the hotel company and was employed in a business and marketing capacity for them. They have confirmed that in an email that I have.

Moving down the age ladder of our patients, we recall memories from Jacob Marshall, who is 25 years old.

Jacob has Sophie's heart and has romantic memories of Sophie meeting and marrying her husband, Zac Coleman. Popular culture associates the heart with romance, but there is no empirical medical or scientific evidence that there is any link.

Jacob states that Sophie met Zac at a Trade Fair in Manchester in 2003. He further states that Sophie almost ended up in Zac's hotel room that first night but got side-tracked by her colleague, who needed help to prepare a presentation to be delivered the following morning. According to Jacob, Sophie did not resist Zac's advances the following evening.

Zac was a man of six feet four, rugged build, square-faced, and looked like he might play rugby if he hadn't been an American. Sophie's time watching rugby at Leicester and her almost reverence for their rugged players may have influenced her feelings for Zac. He was built in the same mould, according to Jacob's description. He says that Sophie had an instant attraction to Zac because of his build, sparkling eyes and disarming smile.

Zac had recently moved to London from the USA

to work in a rival hotel group's office. It seems he was after more international experience in the hospitality industry. He was living in a flat in central London at the time.

There is no evidence to support this claim, except that the Trade Fair did happen. Anything else is just speculation. Considering the clarity of our patient's memories, we might be wise to accept this at face value.

Jacob states that the courtship was intense. He would not elaborate on what he meant by that, just that he felt Sophie was in awe of Zac's personality and confidence. Again, we have no confirmation.

The couple married in Smisby in 2005. Zac's parents were the only members of his family to attend the wedding in the small St James Church in Smisby. The couple honeymooned in Marrakesh because that was where Zac wanted to go. Sophie got the choice of church, so he got the choice of honeymoon venue.

The couple bought a house in Chorleywood as it was close to Sophie's company HQ and on a direct tube line to Zac's company offices in central London.

The following three patients have memories of roughly similar timescales. Interestingly, there appears to be little or no overlap in the content of these memories. This is most peculiar. If the memories of the three cover the same period, I would have expected an overlap. However, I can find only one reference common to all three. They all agree Zac was violent towards Sophie.

Rose Prentice's memories are the less disturbing of the three. She remembers Sophie travelling to Arizona to holiday with Zac in his parent's home. Rose's memories of these holidays seemed to start when she had memories of a place called Sky Harbour. This was a complete mystery to her as she did not sail and, as far as she knew, had never visited any seaside place of that name. Her family discovered that it is the name of the airport in Phoenix. This is correct.

Abigail then described memories of a town called Sedona that is north of Phoenix and is where Zac's parents lived. She was able to report details of the journey, Sedona, their address, and much of the plan of the house.

A telephone call to the Sedona city council offices confirmed what Abigail had said. The address she gave was occupied by a family called Coleman. The description of the house layout was confirmed by their zoning department. They even sent us some photographs of the street and house. They confirm that Abigail's description was very accurate.

Abigail also described Zac's parents as Republican racists. They talked about their governments' inability to stop Mexicans from crossing the border into the State. They thought the Border Patrol should have the right to shoot them while they were still in the Rio Grande River. Their bodies would then be carried away down the river back to Mexico, where they belonged. They were unhappy that their son had married a colonial foreigner.

There is a hint of the violence in the American family in Rose's testimony. Jack Gibson, on the other hand, describes Zac's growing violence towards his wife.

This started about eight months after their marriage, according to Jack. He also mentions that this frequently happened when Zac had consumed too much alcohol. Sophie suspected that Zac was having affairs. She found makeup marks on his shirts of a make she did not use. Sophie also found small items in their bedroom, such as an earring. On one occasion, female panties that were not her size were under the bed. Sophie surmised that he brought women home when she was away travelling for her company.

Jack Gibson got Sophie's liver because he had been a drinker to excess. He was also violent towards his wife and two daughters, which is why his wife left and divorced him. Jack claims not to have been a womaniser, which is how he describes Zac.

Is Jack Gibson's previous lifestyle any indication of why he got similar memories from Sophie? There is no way of knowing. It may also be another coincidence to put alongside Jacob Marshall's love interest memories.

Staying with Jack Gibson's memories, there is one disturbing report from him. He says that Sophie threatened to go to the police about Zac's violence. However, Zac told her that he would kill her if she did. Sophie took this as an idle threat until Zac told her he had already murdered someone, and that was why he was in England and not the USA.

He told Sophie that he and a couple of friends were out drinking one night in Los Angeles. They got very drunk and picked on what Zac described as a bitch, wetback undocumented Mexican. The three beat her to death in an alley and fled. The police did question the three but could not go any further until they gathered more evidence. That was when Zac asked for a transfer to Europe.

The only way to confirm this is to ask the Los Angeles police for a report of the interview. I'm not sure that this exercise would advance our investigation in any way except to confirm Sophie's fear of her husband.

Arthur Morgan's memories take us on to the next part of Sophie's life, the death of her husband in a motor accident. Can we make any connection between Sophie's parent's death, her own accident, and that of her husband, as all three were motor accidents? I do not think we can.

Arthur reports that about two years into their marriage, Sophie got a visit from a lady called Charlotte James. This lady knocked on Sophie's door one evening and was eventually admitted to the house. She said she knew Zac was out and used that opportunity to speak to Sophie. She said she had come to apologise and warn Sophie about her husband.

She told Sophie that she had met Zac at a trade fair. She was smitten by him and accepted his offer to go to his room. In his room, Zac became aggressive and violent towards her, and Charlotte wanted to leave. Zac refused and basically raped Charlotte. Charlotte did not consult the police as she had been seen going to Zac's room of her own free will. It would be a she-said, he-said story.

Arthur states that Sophie was initially angry but soon was sympathetic to Charlotte as she told her story.

Sophie confirmed to Charlotte that she knew of Zac's frequent infidelities and violence. She had found items of clothing and jewellery in her house that were not hers. Sophie showed Charlotte her own bruises from a previous beating and recounted a story of Zac and some friends beating a Mexican immigrant woman to death in Los Angeles.

Charlotte indicated that she would be a witness if Sophie was to bring proceedings against her husband.

Charlotte and Sophie became friendly as a result of their shared fear of Zac and Charlotte's support and sympathy for Sophie's plight.

During a lengthy discussion, Sophie told Charlotte about Zac's claim of killing a woman in Los Angeles. As a result of their shared pain and humiliation, the two decided to stay in touch. They became such close friends that Sophie even registered Charlotte as her next of kin.

Arthur knew that the police had seen Sophie's bruises after Zac had beaten her the previous evening. They had sent her to the hospital to be checked, and a report was filed with the police.

Arthur's final recollection was of the police coming to Sophie's home to inform her of the findings of their inquiries. It was ruled an accidental death while under the influence of alcohol and cocaine, exacerbated by the driver not wearing a seat belt.

In 2008, Sophie had arrived at a point where she could stand Zac's brutality no longer. He returned home one evening drunk and smelling of cheap perfume. He grabbed Sophie and started to punch and kick her about the body. She fought him off and faced him down with a knife. She threatened to kill him if he did not leave. She thought she had managed to nick his arm with the blade to show she was serious before Zac stormed out of the house.

Sophie telephoned Charlotte, who came to help protect her by staying the night. Sophie and Charlotte were both in the house when the police arrived the following morning to tell Sophie that Zac was dead. They reported that he seemed to have been driving very drunk and was not wearing his seatbelt when he went over a steep embankment and smashed into several trees on the way downhill.

Sophie explained what had happened the previous evening and why Charlotte was there. She showed the police her bruises. She also said that Zac often drove when over the limit and frequently did not use his seatbelt. He had points on his licence for one seatbelt incident.

The police investigation, including photographs of Sophie's injuries, was taken for the police at Watford Hospital. The police investigation concluded that Zac died as a result of reckless driving while under the influence of drink and cocaine. His blood alcohol levels were very high, and as a result, he was clearly not in a fit state to drive.

Sophie had Zac cremated and only told his parents by letter after the event. They had belittled her, their son abused her, and she saw no reason to be generous towards them.

The final piece of Sophie Coleman's life was supplied by Megan Thomas. It is a description of the life of a woman who showed a mixture of remorse, guilt and courage after the death of her abusive husband.

According to Megan, Sophie swore off men for a

while. She seemed to have realised why she was drawn to a particular type of individual, just like Zac.

As an attractive woman, she was not short of suitors but continued to decline their offers. Megan said Sophie seemed to feel guilt for her husband's death. There was something niggling at her that Megan could not put her finger on. She was still afraid of something.

At work, Sophie's best friend's young daughter died of heart disease while waiting for a transplant that never came. The family were heartbroken, and Sophie decided to become an organ donor to show her support. She talked to her now long-term friend, Charlotte James. They both signed up immediately. Charlotte also became Sophie's notifiable person. As she had no close relatives left alive, Charlotte became Sophie's next of kin.

Megan's last memory of Sophie's life is of an argument she had with a man called Roberto Caruso in a hotel in Suffolk. I have discovered that this argument happened on the day Sophie died. Sophie's company confirms that she was at a meeting in a company hotel in Suffolk. Roberto Caruso was the manager of that hotel. He is no longer employed by the company.

Graham postulated that as far as he could guess, Sophie's memories could only have left her brain and travelled to her organs in two ways. This was an assumption without evidence.

Memories are held in synapses in the nervous system. Perhaps they were travelling around Sophie's body after the crash and lodged in the organs when brain death occurred. Why this might be the case was beyond

Graham's understanding.

His other thought was that tiny particles of Sophie's neocortex that contained her memory might somehow have made their way through her bloodstream into her organs before the transplant. If that had happened, he could not understand why they fragmented that way and why they stopped there until they were removed from her body. He reminded them that the cornea does not have a blood supply. That probably ruled out that particular idea.

He went on to say that if he was wrong, and that was what happened, there was no way for the reverse process to occur in a recipient's body. The recipient's immune system would have detected these fragments of the foreign neocortex and neurotransmitters and tried to kill them despite the immunosuppressant drugs they were taking. It may be that the fact that this has occurred means that none of the recipients were on powerful enough immunosuppressants.

It would be very unusual for anything like this to happen to one person. Graham thought that for it to occur to eight people would be impossible. As far as he knew, no known mechanism in the human body could enable this to happen.

If I am wrong, wrote Graham, it should have happened before, and it has not.

Graham concluded that he was out of ideas as to how this might have happened. He had reviewed all the information sent to him. He considered every possible anatomical structure he could remember but was no wiser than when he started.

He admitted that he had been clutching at straws,

but there was nothing left to consider, in his opinion.

Graham got Allan to read the history and agreement to forward it to the other consultants.

"I don't know what else you can do," said Allan. "Perhaps the others will have a different view and come up with something. We'll wait and see."

The report was emailed to the other eight that night.

30

Face To Face Meeting

It took several weeks for the seven other consultants to contact Allan and Graham. They all confirmed that their re-examination of all of the tests on their patients before and after the transplant showed nothing they had not seen before. As was usual for transplants, nothing went totally smoothly. There were always problems, but nothing that they had not seen before.

They also confirmed that they had read Graham's paper and could see nothing extra that he might have done to illuminate the problem. They all added their congratulations to Graham on what he had done. Excellent work, even if it proved nothing except they were entering the unknown.

Over a thousand IQ points and a hundred years of experience produced no results. Not a sausage, zilch, nada, rien. None of the consultants could explain how this memory transplant might have occurred. Most agreed that transmission in neurotransmitters or the bloodstream seemed the most obvious answer.

The eight consultants engaged in extensive speculation, discussing various topics, including a potential conspiracy among the eight patients. The details of how this conspiracy might have occurred remained unresolved.

Allan contacted them again and suggested they all get together to decide what to do next. He proposed somewhere on the outskirts of north London, being the easiest for travel plans.

Nicola Stephen replied to the invitation and suggested the Grove Hotel and Spa, just off the M25 near Watford. It was not cheap, but it was nice, had easy access and plenty of parking. They could book a small meeting room for a day and stay for the night.

After numerous discussions via telephone and email, they finally agreed to meet, and Nicola's proposal was deemed the most favourable option. None wanted their Hospital Trusts to know what they were up to. If the Trusts got wind of what they were doing, they might ask questions that none of the consultants were prepared to or even could answer.

Nine rooms were booked for the overnight stay and a small room for the meeting that would be held on a Saturday afternoon. The eight consultants and Graham Bradbury agreed to split the costs equally.

All nine doctors arrived at the hotel just before midday on Saturday, 19 April 2014. They booked into their rooms and agreed to meet at two that afternoon in one of the small boardrooms set aside for their meeting.

Adil Patel and Kasara Doshi met in the bar to grab a snack before the meeting. The two had met on Zoom but not in person. They had the longest journeys from

Cambridge and Oxford and needed refreshment after navigating the M25.

Sitting at the bar to order a snack, Adil spoke to Kasara.

"I made a mistake this morning. I came down the A10, and on to the M25. I must have passed the spot where our strange organ donor had her accident."

"Was that a mistake?"

"It's just that as I was travelling west along the motorway, I had an odd sensation that I was passing near to the spot where she lay. I don't know exactly where it happened, but as I drove I felt a bit unnerved, as if she was watching me."

"That is strange. Is this situation beginning to get to you?"

"I hope not, but maybe it is."

"There has to be a solution to this problem, even if we have not found it yet. No need to get uptight," said Kasara, trying to sound supportive.

"It sort of reminds me of what happened last year with my transplant patients."

"The one that got this lady's organ?"

"Yes, that is the one. When she started to have what she called memories of the donor, I was thrown. We all have transplant patients who have odd dreams, but this woman's became more specific. She is a woman of fifty and I wondered if she had the start of dementia or Alzheimer's. Nothing about what she said sounded as if it could be true."

"That's a reasonable position to take. I did almost the same with my patient," said Kasara.

"It all started to unravel for me when I met Allan

Cunningham at a symposium in Birmingham."

"We were having an idle chat when he started to describe a weird experiences he was having with one of his patient. As he talked, I realised his patient was having the same sort of episodes as mine. I mentioned the donors name and he nearly flipped. We realised we were talking about the same organ donor, we had doubts about our sanity."

"I share your doubts, " said Kasara. "Until I got your call, I wondered about my patient. She is mid-forties, and like you, I wondered about her mental state. Now it appears nine of us are all headed for psychiatric treatment."

Adil raised his cup of coffee in a salute. "Here's to getting some clarity today."

"I'll drink to that," said Kasara, raising his tea.

31

All Talk?

At two o'clock, the nine doctors were all in the small, exquisitely designed meeting room of the Grove Hotel. On a small table sat pots of tea and coffee with cups, saucers, teaspoons, side plates, milk, sugar and artificial sweeteners. There were two large plates with varieties of biscuits.

As the group milled around collecting their drinks and biscuits, there was a round of introductions, and it was nice to meet the person behind the face on the screen greetings.

Allan invited them all to sit down.

He opened the proceedings. "I'd like to introduce Graham Bradbury." He tapped Graham on the shoulder, who was sitting beside him. "We have all read his report and I'd like him to open the proceedings if you all agree?"

There was a nodding of heads and a chorus of yeses. Graham did not stand up to speak.

"Thank you, Allan, and thank you all for the information that you sent me. As you can imagine, it has been a mammoth task to sift through everything.

I have been fortunate in that a few members of our surgical team who knew what was going on, helped with the work."

"The first question I address is whether this memories transplanted into our patients is real or imagined."

"Much as I hate to say it, I think that what our patients are experiencing is real, and not a figment of their imagination. The evidence for this is that all eight patients independently gave the name of the donor depending on what part of her memory they appeared to have. My assessment is based on nothing more substantial than that at the moment."

"If their memory concerned the donor's life before her marriage, our patients name her as Sophie Beckinsale. Patients remembering post-marriage events call her Sophie Coleman. This is too much of a coincidence to ignore."

"There was an initial thought that this might be an elaborate scam or hoax. They might have all colluded in some way and collected information from the internet, or any other open source."

"For eight, seemingly random people, to somehow get these particular details from the internet would involve a very elaborate scam. If it is a scam or hoax, I see no rationale for it. What might such a hoax achieve in the long term, except eight hospital departments with egg on their face? Not really much of an achievement for so much effort."

"Equally, they would all need to be known to one another, and there is no evidence that this is the case. They certainly could not have set it up prior to their transplant, as none knew that they would get a new organ, let alone all eight from the same donor."

"Setting something like this up after their operations is just about possible, but as I said, to what end? Not only that, but some of our patients were having a hard time for the first few months after their surgery. They were struggling to get used to their new situation, and all of the emotional trauma that accompanied it? I doubt many of them had time for other nefarious pursuits."

"Looking at Sophie Coleman's life, there is much about it that might have been gleaned from open sources. The first issue for our patients would be to identify the name of their actual donor. One person might have done this, but how would they know who to share it with. The names of the donor and the recipients are kept very confidential in Bristol."

"Let us assume for the moment that the donor's name had been discovered by one of our patients. She could have lived anywhere in the country. How would they know where to start to look for information? It would have taken a great deal of effort and ingenuity to know where to look and what to ask. I know, because I had a great deal of information about her, and I still struggled."

"What is troublesome to me is that our patients have given us details about her life that could not be gained this way. For example, our own kidney transplant patient challenged us to talk to Sophie's friend from De Montford to confirm his story. He named her as Joan Tapp and said she was a friend of Sophie who lived in the same block of flats. He even claimed that she was studying chemistry."

"I contacted De Montford and they confirmed our

patient's story. Joan Tapp studied chemistry and lived in the same student accommodation as Sophie at the same time. They could not confirm that they were friends, but the university suggested that since they were in adjoining rooms, and shared some domestic facilities, that was likely. As a result of this, and the many other pieces of information that I have received from you all, I have concluded that our patient's situations are real."

"Mrs Mason." Graham looked towards the lady.

"Please call me Isabel."

"Thank you. As a starting example, your patient remembers Sophie's holidays in Arizona. It's where her in-laws lived. She described the street and the house in a town called Sedona."

"I contacted the local police and municipal authorities in that town. They confirm that a family named Coleman were registered in the street described by your patient. The local authority sent me a floor plan and photograph of their house. Your patient could have been describing the drawing and photographs they sent me."

"Your patient has also more recently talked about travelling along a stretch of the famous Route 66 on her way to the Grand Canyon during one of the two holidays she took in Sedona."

"To get from Sedona to the Grand Canyon, Sophie would have travelled through Flagstaff. That piece of road is still sign posted as Route 66. There are photographs on the internet showing the iconic sign in the town. I have no knowledge if Sophie got some kicks from her violent husband on Route 66, but she will have travelled that road going to the Grand Canyon."

"Professor Khan, your patient can recall the full police investigation into the death of Sophie's husband in a car accident. It is a true reflection of the police report they sent to me."

"Mr Isaac, your patient recalls the violence of Sophie's husband towards her. Your patient is the only one who appears to have a similar personal experience to the one he remembers about the donor. One out of eight is not good odds. And so, it goes on."

"As a result of these, and other examples, I have regretfully concluded that our patient's situations are real, but I'm damned if I know how it happened."

"As a man of medicine and science, I thought that there might only be two possible routes."

"As you are all aware, memory is held in various parts of the human brain. I will not go into any detail about how and where they are formed and stored. Suffice to say that we believe that neurons, chemical neurotransmitters and synapses play a vital role."

"I'm going to digress a little here and talk about blood. I considered a blood transfer of parts of Sophie's neocortex containing memory to her organs, but the cornea has no blood supply. This fact alone seem to preclude that particular avenue."

"If I am wrong, and this was the mechanism in the donor, how did the reverse transfer happen in our patients. A particle of neocortex carrying a patient's memory to the pancreas is unbelievable, but the journey from that organ back along the blood vessels, into our patient's brains is even more difficult to understand."

"I'll return to the nervous system now. I said earlier that neurons are the bedrock of memory, and perhaps they are. However, recent research suggests that non-neurons also have some cognitive abilities such as might be found in plants. I wonder if his plants remember what Prince Charles said to them?"

This brought a laugh and a welcome bit of light relief.

"The key part of the nervous system, as far as memory is concerned, are the billions of synapses. Memory is transmitted along the chain of nerves, with chemical neurotransmitters bridging the gap between each nerve fibre. Perhaps some of these chemicals somehow got into the blood stream and ended up in Sophie's organs. Perhaps some of the adjacent nerve fibres themselves were broken off in her violent crash and transferred in this way. I just do not know."

"To conclude, I come back to my basic problem. If memory can be transferred from the brain to any organ in the human body, how does it make the reverse journey from the donated organ into the recipient's brain?"

"The whole thing is insane." Graham stopped, and the room was silent for a few moments.

"I think we all might have to have our sanity checked," said Isabel Mason. "Perhaps we could save money by employing the same therapist."

"Friedrich Nietzsche, had said that: Insanity in individuals is something rare - but in groups, parties, nations, and epochs, it is the rule. Perhaps we can use this to negotiate a group rate for our therapy," said Keith Isaac

"I once read a quote that said the distance between insanity and genius is measured only by success. Now all we need is success to prove our sanity," said Mohamed Khan.

The use of humour did not mask their understanding of the actions they were taking. Still, if they found an answer to this problem, they would be clinical heroes, not clinically insane.

Giuseppe Da Silva was next to offer an opinion.

"I think that Graham is probably correct in that these memories are real, at least to the patients. He discussed the nerves and blood as possible routes of transfer from Sophie's brain to her organs. He is also correct in reminding us that the cornea has no blood supply. That alone would tend to rule out blood as an option. If, as Graham suggests, we are wrong about this, then how did the particles of neocortex containing Sophie's memory make their way back to the patient's brain, and not some other part of their body?"

Graham spoke next.

"As part of my research, I dug into the various types of memory that humans have. I have notes here." He indicated his iPad.

"Our types of memory are: episodic, semantic, procedural, working, sensory and prospective. I must confess that I had to look this up, it's been a long time since I did any of this."

"The ones that seems to fit our problem are prospective and semantic memory. Prospective memory is a sort of mental time machine, in that it allows the person to go back to a particular moment. Semantic memory relates to relationships between objects or concepts. It also allows us to recall our own life and behaviours."

"Prospective memory is held in different parts of the brain, particularly the hippocampus, but other areas as well. Sematic memories are ultimately thought to be stored throughout the neocortex. Given the recent research about non-neuron memory, it could all be rubbish."

"During her accident, Sophie's head was thrown about quite a lot, as witnessed by her basal skull fracture, and several countercoup brain injuries. Perhaps this is where the fragmented pieces of her memory became dislodged from her hippocampus and neocortex. It is a theory about how her body came to have tiny fragments of brain tissue that might have contained her memory. How these fragments got to her organs is a mystery?"

"You have looked very carefully at Sophie Colman's life as written in your excellent report," said Isabel Mason. "We have all given you as much information as we could about our patients. From what you have written, there is no clue in any of their lives as to why this could have happened. It would be extremely strange, or bizarre, if eight random people, and only those eight, had exactly the same genetic ability to accept and decode memory from a tiny fragment of someone else's brain tissue. It is more sensible for us to look at the donor in more detail. She is the only common factor in this equation. How we might get more than what Graham has uncovered, is beyond me at the moment."

"It is difficult to imagine memories travelling around a patient's body as particles of neocortex or chemical neurotransmitters." said Allan Cunningham. "It's memory that we use to remind us how to walk and talk. The message is sent unconsciously to the place it is required to work. As far as I am aware, it does not wander about the body aimlessly checking every home with a door to which it has a only key. I don't see memory as a nomad looking for an oasis."

"I cannot see tiny pieces of neocortex or hippocampus with specific memories, swimming about in a human body, not even after a severe crash," said Keith Isaac. "I think the blood route is not practical and not just because the cornea has no blood supply. Between us we have years of experience and what I am hearing here today is that we don't have a clue. Perhaps the solution to our problem lies with another speciality, such as Neurophysiology. As we are all aware, this is a speciality that deals with functions of the nervous system rather than its architecture. I think that any solution to our problem might come from someone like that."

"A short while ago, I read an interesting paper on this subject in preparation for this meeting," said Adil. "The article started by reminding the reader that how the human brain works is still a very large question. There was much talk of using functional magnetic resonance imaging (fMRI), EEG/MEG and PET scans to try to get a better understanding of the relationship between the anatomical and functional interactions."

"The conclusion was, that even with all of these highly sophisticated instruments, and specialist minds working on the subject, we are still a very long way from understanding. The best the paper could offer was that we keep looking. As far as I can tell, we medical practitioners are still not able to define brain health in a way that is universally recognised. I guess that what I am saying here is that we keep looking."

"If I may?" Said Mohamed Khan in his deep, strong voice. Everyone turned to him.

"I am of the opinion that the blood, as a route of

transport for pieces of memory, is not viable. You might expect me, the only surgeon who transplanted and organ with no blood supply, to say that. While that is true to some extent, there are far too many variables to make it viable."

"How big a piece of hippocampus or neocortex does it take to contain the parts of Sophie Coleman's memory that our patients appear to have? It may be that a tiny particle of either can contain a memory. On the other hand, it might require a lot, microscopically speaking."

"The issue with the blood as a vehicle of transport falls down when one considers the return journey these particles would have to make in your patients, not mine. How would the recipient's body know what to do with foreign matter apart from trying to kill it? It is hardly likely to signpost a route from the organ to the brain and wave it along with a hearty, this is where you belong, and need to be, be on your way. It is my opinion that the only realistic route of travel for these memories is via the nervous system. The cornea is one of the most densely innervated and sensitive tissues in the body. If my patient was going to get memories from a transplanted organ, they would have to have travelled along the nerves. There is no other way."

There appeared to be general agreement among the others when Allan intervened.

"Perhaps we should take a comfort break, and I'll organise more drinks. Return in about a quarter of an hour."

32

Inconclusive Conclusion

After their comfort break, the nine poured more drinks to fill the space left. They sat down to continue their deliberations.

"It appears that we are all agreed that the most likely route of transfer of these memories is along the nervous system," said Allan. "Are we all agreed that transfer via the blood stream is more unlikely than the nervous system.? Does anyone have any insight as to how that might have happened?"

"The only conceivable way is by some form of chemical neurotransmitter transfer," said Keith. "Somehow, chemicals that were already coded by our donor moved from the brain into her organs. We transplanted those organs into new bodies. What happened next is anyone's guess. I find it impossible to understand why those particular coded memories were lodged in any of the donor's organs, and yet, if we believe our patients, that must have happened."

"For these memories to be recalled in our patients, their brains must have accepted the chemical transmitters, decoded them, then allowed them to form new pathways. This is the only way that I can see how our patients can recall some of Sophie Coleman's memories," said Allan. "I know of no way in which this could occur. There is no record of anything like this happening in the past."

"None of us are specialist neurologists or neurophysiologists," said Nicola Stephens. "This is completely out with our skill sets. I suspect that if we are ever going to get to the bottom of how this occurred, we are going to have to find a specialist to help us."

"If we do that," said Keith Isaac. "We are starting to lose control. I have no doubt that such a specialist would want to conduct any number of fMRI, EEG, MEG and PET scans on all of our patients. What do we tell them if we ask them to participate? What if they refuse?"

"What if we set one of these specialists a hypothetical puzzle and ask them to give us their best assessment of how it might happen," said Isabel. "It's a way of not involving our patients directly and might give us some clue as to where we might look next?"

"Just think," said Graham. "If you crack this puzzle, you might all be in line for Nobel Prizes."

"If only it were that easy," said Kasara. "This jigsaw puzzle seems to me to be a made up of a trillion pieces, with no picture of how it should all fit together."

"I know of a woman at the John Radcliffe in Oxford who specialises in neurophysiology," said Adil. "Her name is Holly Barron. I could approach her and try to get some advice."

"That might be the best idea," said Allan. "Meanwhile, are we all agreed that we need to keep control of this for the sake of transplant surgery all over the world?"

"I'm more concerned with how my Hospital Trust Board might react if they had this information," said Giuseppe. "I hate to think what they might do for a bit of cheap publicity."

"I suspect we are all in the same position," said Nicola. "Anything that might get them some publicity and more funds would be a Godsend to my lot."

"If this did get out into the media," said Mohamed. "Think of the damage it would do to our patients. They might well have dozens of journalists camped on their doorstep. The religious lunatic fringe might become obsessed with them. As we are all aware, transplant patients can have significant emotional difficulties in the months and even years after their operations. The inevitable media attention could well make that much worse. They might inadvertently default on their medication, and that could cause serious harm to them, and problems for us as their clinicians. These issue could lead to their early death. Some might even be driven to commit suicide. All of these things would be unforgiveable."

"I think all of these things are very relevant," said Allan. "We need to keep this under wraps if we possibly can for the sake of both our patients in particular, and transplantation in general. We need to have a co-ordinated approach as to how we calm our patients reasonable anxieties. Any suggestions?"

"I've had a little experience of something like this in my past practice," said Giuseppe. "Hallucinations, or in our cases, memories are internally-generated sensory experiences. Hallucinations are more common than we might care to believe. Around twenty-five per cent of the population experience them at some time in their lives. We are all well aware that transplant patients experience them fairly frequently."

"Asking direct pointed questions often raises the individuals hackles. Better to use a more conversational approach. If our patient is distraught, one approach might be to reassure the patient that hearing voices is quite common among a large number of people. We could impress upon them that this is very often true of transplant patients. We could remind them that we see this sort of thing fairly regularly. We don't know how it happens, we just know that it does. Try to get them to talk to one of your hospitals qualified and trusted talking therapists. Treat it as something that is not unusual and entirely manageable. Whatever we do, we must, I believe, lead them to understand that they are not unique, special or crazy."

"I remember a conversation at a previous meeting that suggested making our patients feel that they are the only ones having these dreams or memories," said Isabel. "Now we are suggesting that what they are experiencing is quite common among transplant patients. I'm confused, and maybe that is the way to keep our patients wrong footed. Hallucinations are common in transplant patients, even dreams of the donors life. Nothing unusual in what they are seeing or imagining. Make a good fairy tale for the children or grandchildren. Spreading confusion is an underhanded tactic that might work."

"Has it occurred to anyone that something like this might have happened in the past?" Said Adil.

"Surely we would have heard about it," said Graham.

"Why?" responded Adil. "We are frantically trying to work out ways to hide this for the sake of transplantation in the future. What if other groups of surgeons have encountered the same issue and come up with the same solutions as we seem to have reached?"

"Great minds might think alike," said Kasara.

"We were discussing our sanity earlier," said Graham. "Perhaps it's more a case of fools seldom differ."

"I think that if it had happened in the past, there would have been some leakage. Even if was only a vague rumour," said Nicola. "I'm guessing that none of us has even heard the faintest whisper."

"I think if any of us had heard," said Giuseppe. "we might have mentioned it before now."

It was clear to all that they were no nearer to understanding how this might have happened. There were a few more comments that added nothing to the solution.

"Let's sum up before we break for dinner," said Allan. "We are discounting the blood as a method of transfer of particles of brain matter carrying memory to other organs in the body. The only realistic way for transfer to occur would appear to be along the nervous system. How that might happen we do not know."

"Our biggest issue, as I see it, is the reverse transfer of this neurological information back to the organ recipients brains, for them to recall memories of a now dead person."

"Keeping a tight grip on the release of information by our patients or any of our hospital staff is key. If this gets out, it could be catastrophic for transplantation.

We have no way of knowing how many people might be affected by the publicity. If this has happened in the past without our knowledge, then the clinicians involved have been very successful in keeping it secret."

"The only avenue of investigation that we have identified is to seek some discrete advice from Kasara's contact at the John Radcliffe in Oxford. Holly Barron I believe you said, Kasara."

"I did. I've met her a couple of times and I'm sure she will be discrete. Where she is able to offer any insight as to how this occurred, I don't know."

"Is there anything else that I might have forgotten?" Asked Allan. There was a few shaking heads, but no one spoke. We have a table booked for eight in the restaurant. I'll see you all there."

At dinner that evening, the talk at the table started on the subject they had been discussing all afternoon but quickly fizzled out. It soon became a general discussion about the state of the NHS, especially their own Trusts.

Graham looked at the group and wondered if the hotel had enough seltzer to cure all the bellyaching around the table. Perhaps I'll end up like them one day when I become a consultant, he thought.

33

One Down

The pain was real. He could feel it on his arms, ribs and abdomen. It was a furious attack driven by drink. The squeals and sobbing he heard were those of a woman who was being beaten by a man. The sound in his head was female, but the pain was felt in his masculine body.

Jack Gibson was suddenly awake in a blind panic. Sitting upright in his bed, he felt and heard his heart pounding. Jack was sweating, trembling and gasping for breath. He wondered if he was dying.

"Oh, God! What the fuck is going on?" It was an angry statement rather than a question. There was no one else in the pitch-black room who could give him an answer.

Still trembling, he switched on his bedside lamp. The clock on the same table showed one forty. He flung his legs over the edge of the bed and tried to compose himself. He sat with his chin in his hands and elbows on his knees for several minutes, attempting to understand what had happened. His breathing was rapid and shallow. He was afraid, but he did not know what he feared.

As he sat there, his eyes darted around the room, half

expecting the man who had administered the beating to be there. He was not. He was alone in his tiny bedroom.

As his heart rate dropped and his breathing started to return to normal, he remembered that this was not the first time he had had this dream.

It started about eight months after his liver transplant operation. He had been having memories of a woman who had donated his liver, and they were becoming more detailed.

He started to have memories of her husband beating her. He did this when he had had too much to drink. It was not just physical abuse. He verbally abused and belittled her. Jack thought he was a nasty piece of work.

Tonight, it was different. Even after composing himself, Jack could still feel the pain of the beating. Feeling the pain of the beating was a relatively new experience for him.

As he sat there wondering why this was happening, Jack began to remember why his wife had left him and taken their children. He had been physically abusive towards her and verbally abused his young children. He remembered shouting at them and could see them frightened by his anger.

If his organ donor's husband was a nasty piece of work, what did that make him? He was no better. It had taken some time, but he had apologised to everyone concerned. He had apologised because he was told by his therapist that it was wrong and not because he felt any guilt, not until this particular night. Tonight, Jack began to feel guilty for what he had done to his family.

This woman's memories had put him in the place of his wife. Almost for the first time, Jack realised what he had done. The shame and guilt were as palpable as the pain he had felt during his dream and after he woke up. For the first time, a tear of shame ran down his cheek.

Jack had always been able to blame someone or something else for his shortcomings. It was his army service and being shot at that made him drink. His wife was argumentative, and his children were cheeky; they all needed discipline.

This night and this dream had given Jack his first real insight into his behaviour towards his family. It was not a dramatic road to Damascus moment, but it marked a beginning.

Since his transplant operation, Jack's health had improved, but his social circumstances had not. He found it difficult to get a job, and when he did, he had difficulty keeping it. He was well behind on his child support payments, so far behind that his wife had given up chasing him. He had to move into increasingly cheap accommodation, which he could only afford with his benefits.

He had only spent twelve years in the army. That did not qualify him for a pension until he reached the usual retirement age. He had not claimed a War Disability Pension for his PTSD because he did not know that he could, and no one made him any the wiser.

Unfortunately for Jack, the dreams of beatings of his donor became more frequent. He was struggling to handle them. He talked about them to his principal transplant coordinators. She listened and gave what advice she could within the limits of her professional training. She got Jack a series of appointments with one of the King's College Hospital specialist psychologists. The results of these meetings were passed to Keith Isaac and his team. Keith wondered if Jack might return to drink. He made his worries known to the psychologists and cautioned her to be aware of the signs.

The meetings gave Jack some respite, but nowhere near enough as it turned out.

One night, about eighteen months after his operation, Jack had had an especially tough day. He was alone at home brooding over the memory he had that afternoon of the beatings that Zac Coleman had given his wife. He had felt Sophie's pain as Zac rained down blows on her body.

In all that pain, he remembered what he had done to his wife. He felt what his wife must have felt. Why had he done that to her? He married her because he loved her. They had two lovely children, and he had verbally abused them. It was not their fault. He was the only one to blame. Jack was in the depths of despair.

He left his bedsit and walked a few hundred yards to his nearest off licence. There, he purchased the cheapest bottle of vodka they had. It was all he could afford on his benefits money.

Back in his bedsit, he sat in his chair and slowly drank from the bottle while berating himself for being the shit that he was. The more he talked to himself about how horrible he had been, the more he drank and the worse his mood became.

It took some time for Jack to reach the end of the litre bottle of neat vodka. By the time he reached the bottom of the bottle, Jack was raging at himself for what he had become and how he had treated his family. It was a foul-mouthed rant.

He also raged and cried in support of Sophie Coleman. His memories of her indicated that she, like his wife, was also an innocent victim of a drunken, violent slob. He was cursing and shouting as he staggered around his tiny bedsit.

In his drunken rage, Jack swore that if he could, he would find this Zac Coleman and kill him. He did not deserve to live.

"I'll kill that lousy bastard if I ever meet him. He does not deserve to live."

"If he does not deserve to live, why do I? We're both the same. Sophie Coleman did not deserve it. Neither did Amy We're the scum of the earth."

With that, Jack stumbled into his little kitchenette and grabbed his carving knife.

There was no Shakespearean soliloquy about seeing a dagger before him or even any rational thought regarding what he was about to do. It seemed like a good idea at the time. He placed both hands around the handle of the carving knife. Holding the blade towards his body, he plunged it into his chest. He died instantly.

Jack's body was found by his friend Terry Brogan, who had called to see him the day following his suicide. He got no answer to his knock, and when he looked through the letterbox, he saw Jack's body lying in his kitchenette, covered in blood. Terry called the police, who broke down the door.

As it was obviously a violent death, it was classed as suspicious. The investigating police officers made their minds up within a day. There was no forced entry before the police broke down the door. Only Jack's fingerprints were on the knife handle. The deceased blood alcohol level was almost off the scale, and there was an empty bottle of cheap vodka in the kitchenette. Suicide, no doubt, but they had to go through the motions for the coroner.

The pathologist report gave the cause of death as a single knife blow penetrating the body below the sternum. The blade had penetrated the lower part of the left ventricle of the heart. It travelled upward through the wall between the left and right ventricles before exiting the heart through the upper part of the right ventricle.

The pathologist did not add his thought that if it was a suicide, it was an extremely accurate stab, given the level of alcohol in the blood.

Keith Isaac learned of Jack's suspected suicide two days after the event when the police came to speak to him. They said it looked like it might be suicide, but if it was, it was the strangest way to kill himself that they had seen. The deceased had left no note, which was also odd. They had noted his medicines and scars on his chest, which was why they wanted to speak to Keith. They also informed Keith that he had a very high blood alcohol level. There was an empty one-litre bottle of vodka in his sitting room. It looked like he had drunk the lot himself. Could Keith shed any light on why he might do it? He could not.

He was shocked but not entirely surprised, Keith had said. Mr Gibson had been a heavy drinker for many years, and that gave him significant liver disease. Keith told the police he had refused to perform the liver transplant unless Mr Gibson gave up the drink. He had and seemed to be coping well. Mr Gibson was addicted to alcohol. Even if he never touched another drop, Mr Gibson would always be addicted to alcohol. He left the comment hanging for the police to draw their own conclusions.

He asked the police to return in two days when his team's psychologist would be available. She had spent some time with Mr Gibson trying to resolve some emotional issues he had. She would be the best person to talk to. They agreed.

Keith texted Gina Ramage to contact him urgently when she returned from her short break. Jack Gibson seems to have committed suicide, he wrote, and the police are asking questions about his mental health. Bring all his notes when you get back.

Gina had recorded how Jack Gibson was being plagued by the memories of the beatings that Zac Coleman had visited upon his wife. Feeling Sophie's pain was a relatively new problem for Jack.

Gina, like everyone else, struggled to understand how a piece of the donor's memory could be transplanted with their organ. Worse still, memory transfer was one thing. Feeling the pain of the beatings as part of that memory was quite another.

She also noted that Jack was increasingly agitated about how he treated his wife and children. It seemed to her that Jack's memories of violence towards the organ donor were bringing into sharp focus the memories of his own domestic violence.

In her last report, she thought that the memories of the donor might enable her to help Jack come to terms with his own memories of domestic violence. She was eager to investigate this issue at their upcoming meeting.

Jack's suspected suicide meant that would never happen.

Keith's worried that he and his team might not have recognised Jack's state of mind earlier. How culpable were they in allowing this to happen?

As it appeared that Jack left no note, there was bound to be a coroner's inquest. Suicide could not be taken for granted if the death were deemed to be violent. Suicide had to be proved, not assumed. An added complication for Keith Isaac was that Jack was under his care in hospital. The coroner would want to be satisfied that his care did not contribute towards his death. Given Jack's personal history and what had been happening with his memories of the donor, this could be a problem for Keith and the hospital.

What was certain the hospital, Keith's Isaac and his staff would be asked awkward questions at the inquest. Keith hastily called a team meeting. They had to get their ducks in a row.

When his staff had assembled in his office, Keith outlined what had happened to Jack Gibson. It was thought that he had committed suicide with a knife through his heart. This was an unusual way to commit suicide and was bound to be investigated. As he was under the ongoing care of Keith, he and his staff were bound to be asked how this might have happened.

"We are all aware of what has been going on with Jack Gibson, and the so-called memories of his organ donor," said Keith. "You are also aware of what might happen if any hint of memory being transplanted makes the media, in all its forms. This tragic event must not provide a conduit to anyone outside this room gaining access to our findings on this subject. It must be contained. Gina, can you summarise your notes on Jack's emotional state?"

Gina, the psychologist, stated that Jack was having difficulty reconciling his own violent past with his dreams of violence towards his organ donor. She thought that they were beginning to get to grips with the issue.

"We all know that Jack Gibson thought these dreams were the actual memories of his organ donor," said Keith. "Those in this room know where we are with this problem and how it might affect the future of transplantation, worldwide if it got into the media. I'm sorry if I am labouring this point."

"Having given this some thought since I learned of Jack's death, I think we must ensure that the information we present shows that Jack's fixation with his donor is based on dreams or hallucinations, not real memories."

"Any suggestion in our notes of real memories must not come to light. This is not a lie. We have no way of knowing if Jack had real memories of his donor, or if they were just dreams. There is no known medical route for memories to be transplanted with an organ. Without that evidence, Jack Gibson had dreams and nightmares, not memories."

"I will take the lead on this when it happens. I will not be telling a lie if I state that Jack was having dreams and nightmares. Gina, I believe that your notes talk about Jack's dreams of his donor are remarkably similar to his actual experiences with his wife. Am I correct"

"Yes. I was working on the premise that Jack's psychological problems relating to his organ transplant stemmed from his memories of his own violence towards his wife and children. There is nothing in my notes to suggest anything other than historical family influence on his mental wellbeing. Given that he might have taken his own life, I cannot see his action has any bearing on the fact that he had an organ donations. This was a procedure that gave him a chance of a new life. Perhaps the generosity of an unknown organ donor made him more aware of how selfish he had been towards his own family? We will never know."

"That's a good point, Gina. Can you all please ensure that whatever you have written or heard on the subject of Jack Gibson having real memories never sees the light of day if there is an official enquiry. I'm not sure it will come to that but we must be prepared. Are we all agreed?"

They were.

34

Coroner's Inquest

Jack Gibson's death was unexpected, unnatural and violent. Those facts required the police to investigate and report it to the coroner. As soon as the death was reported, a coroner's clerk started gathering evidence.

The coroner has four questions to answer:
1. The identity of the deceased.
2. Where the person died?
3. When the person died?
4. By what means did they die?

The coroner has no power to investigate why the person died. That is for the police to investigate.

The police gathered evidence at the scene. The forensic team bagged the knife, the empty vodka bottle, and Jack's bottles and sleeves of pills. They photographed everything because it was a violent, suspicious death. A cursory look at the knife sticking out of the deceased's chest would instantly suggest murder.

After a short time at the scene, the police could find no evidence of forced entry before they arrived on the scene. They had smashed down the front door, but all the windows in the small flat were intact.

They started to consider the death was likely suicide but could not proceed based on such an assumption. The one serious flaw in their assumption was the lack of any suicide note.

The standard of proof required for coronial conclusions is ordinarily the balance of probabilities. For a suicide conclusion, the coroner must establish that the deceased committed an act that resulted in their death and, importantly, that they intended that act to result in their death. Suicide was never to be presumed and always had to be established beyond a reasonable doubt – that is, to the criminal standard of proof. The police had to investigate accordingly.

The police noted that the deceased had a large healed scar on his chest from a recent operation. His home had many boxes of pills. Was there a medical aspect to the deceased's death? The knife sticking out of Jack's chest seemed to confirm the cause of death and that it was violent. However, were the drugs he was taking in any way complicit. The coroner ordered a post-mortem.

The pathologist established that the cause of death was the eight-inch carving knife that had punctured the deceased's heart. It had entered his chest below the sternum and travelled upwards. It passed through the left ventricle, the dividing wall between the left and right ventricle, before exiting through the upper part of the right ventricle. It was a powerful thrust.

The pathologist reported that the deceased had a blood alcohol level equivalent to him having drunk the whole one-litre bottle of 40% vodka. That did not kill him but would severely impair his ability to reason logically.

There were no defensive wounds on the deceased. The pathologist indicated the existence of the transplanted liver and the healed scars from that operation.

The drug toxicology report had not been received due to the numerous bottles of medicine discovered in the house.

The police only found the deceased's fingerprints on the knife handle and no sign of forced entry other than their own. They had forced entry in response to an emergency situation.

An inquest was hastily arranged and adjourned to allow the police time to gather the evidence.

The police contacted Amy Gibson, Jack's divorced wife, by telephone. They reported his death and asked if she would come to London to officially identify the body.

"Are you sure it's him?" She asked.

"We are."

"Good. Now I won't be lying when I tell my kids their dad is dead."

Taken aback by this outburst, the police asked her to explain. She outlined her ex-husband's drinking and violence towards her and the children. That was the reason she left and divorced him. He was always behind with child support, and she was glad he was gone. She'd have no reason to speak to him now. She did not want to see him either.

The toxicology report was completed. It found levels

of the medicines in Jack's blood consistent with the treatment his transplant team indicated he should be taking.

Keith Isaac confirmed the drugs and added that his patient was undergoing psychological treatment for disturbing dreams and hallucinations. Dreams and hallucinations were not uncommon in transplant patients, but neither did they affect all such patients.

A date was set for the full Inquest.

Mrs Shirley Fleming, the coroner, opened the proceedings. She called Terry Brogan as her first witness. Terry was sworn in and gave his account of finding Jack Gibson's body.

"Why did you visit Mr Gibson's home that morning?" Asked the coroner.

"I had met him a couple of days earlier. He was feeling down and was upset. I thought I'd check to see how he was getting on," said Terry.

"Did Mr Gibson tell you why he was feeling this way?"

"He told me he was having bad dreams about his organ donor. He told me she was a woman who was beaten by her husband. His wife divorced him for that same reason and he was feeling guilty. He said he could feel the beatings."

"Did he say that he felt the beatings?"

"He did."

"Who was the subject of these beatings?"

"He said it was his liver donor who was being beaten."

"Did he know the donor?"

"He said he met her in his dreams and memories."

"I'm sorry. Did you say he met her in his dreams and memories?"

"That's what he told me."

"So, what you are saying is that Mr Gibson told you that he met his organ donor in his dreams, and that she was a woman who was being beaten by her husband. That seems a very unlikely story."

"That's what I thought, but that's what he told me. I thought he might be going a bit doolally." Terry tapped the side of his head as he spoke.

"You say that his wife left him because he had been violent towards her."

"That's what he told me. He wondered if him feeling these beatings was some sort of mental revenge for his actions against her."

"Did Mr Gibson ever suggest that he would take his own life?"

"No. He just said that he felt guilty for what he had done to his wife. He blamed it on his drinking."

"Did you see him drinking?"

"No. I only met him after his liver transplant. He'd given up the drink then. He told me he had needed the transplant because he was an alcoholic."

"Is there anything else you would like to add, Mr Brogan?"

"No."

"Thank you, Mr Brogan. You may step down."

The next witness was the pathologist who had performed the post-mortem. His evidence confirmed his written report as to the cause of death. He confirmed the high level of alcohol in the deceased's blood and the finding for drugs other than those prescribed by the hospital.

He confirmed that there were no defensive marks on the deceased.

"This does not prove that he was not attacked by an assailant, does it?"

"No Ma'am. I might have expected defensive marks as the wound appears to have been delivered from in front of Mr Gibson. However, the level of alcohol in his blood will have severely impaired any reaction to impending life threatening danger."

"What does the wound tell us about a possible attacker?"

"The entry of the knife suggests that it could have been delivered by a right-handed man stabbing from the front or rear, or a left-handed murderer stabbing from the front."

"Were any fingerprints found on the knife?"

"There were, Ma'am. Only the deceased's fingerprints were found."

"Would an assailant wearing gloves have erased these fingerprints?"

"Possibly or smudged them, and that was not the case."

"Indeed. Were the levels of any of the prescription drugs excessive?"

"No Ma'am, they were what would be expected given his drug taking timetable."

Next up was the police in the burly shape of Sergeant Tony Batters.

He explained why they had broken into the flat

without a warrant. They saw a body covered in blood lying on the floor, making it an emergency situation.

They could find no evidence of forced entry other than their own. All of the windows were intact.

He confirmed the details of their investigation, including the report that only the deceased's fingerprints were on the knife.

Despite an exhaustive search, they could find no suicide note. The formal identification of the body was completed by Mr Brogan. Mr Gibson's wife declined their invitation and expressed no remorse for her ex-husband's death. Sergeant Batters outlined the telephone conversation he had with Amy Gibson.

"That would seem to confirm the evidence given by Mr Brogan," said Mrs Fleming.

"It would, Ma'am."

Keith Isaac was next in the witness box. He confirmed the details of the liver transplant noted by the pathologist.

"You have heard Mr Brogan describe Mr Gibson's disturbing dreams of his donor. What can you add to his testimony?"

"It is not unusual for organ recipients to have dreams and nightmares about their donor. Most commonly, it's remorse that someone died to give them an organ and a better life. Many patients have vivid dreams in the recovery room after operation. If they are placed in Intensive Care for any length of time after their operation, they frequently have dreams, hallucinations, memory loss, and confusion."

"Was Mr Gibson in Intensive Care."

"He spent four days there before transfer to the main transplant ward."

"Did he have any of the effects that you have outlined?"

"His notes indicate that he had them all, frequently."

"His friend talks about his recent dreams and memories. What do you know of that?"

"We were well aware of how he was feeling. I sent him to the clinical psychologist assigned to my team, Ms Gina Ramage. She was treating him for his demons when he died."

"Have you anything further to add, Mr Isaac?"

"Yes Ma'am. When Mr Gibson was initially referred to me by his GP he was an alcoholic. I also discovered that he had seen active service with the army in Iraq and Northern Ireland. I suspected he had undiagnosed PTSD. I sent him to a specialist who had good result with Mr Gibson."

"I also initially refused to put Mr Gibson on the transplant register until he gave up drinking for the rest of his life. I gave him the option of allowing me to keep him comfortable as he drank himself to death, or stop the drink and get a chance of a new life with a transplant. He chose the latter option. I was very disappointed to learn that he was full of alcohol when he died."

"Thank you, Mr Isaac."

Gina Ramage took the stand. She confirmed that she had consulted with Mr Gibson on five occasions, before his death.

Gina reported that she thought Mr Gibson was becoming even more remorseful and withdrawn because of his past violent behaviour towards his own family.

She wondered if he was projecting these feelings onto an unknown organ donor.

She thought they were beginning to get to the root of the problem at their last meeting before he died. She confirmed that he had expressed his desire to sort out his issues. She had seen no indication that he had taken up drink again.

The corners summing up was short and sweet, as far as the hospital staff were concerned.

"On the basis of the evidence presented here, there is no evidence of Mr Gibson being attacked by another person. There is also no note or proof that Mr Gibson intended to take his own life although that might be possible."

"Consequently, I am recording an open verdict. Thank you all for your time."

Case closed.

Keith Isaac and his team heaved an enormous sigh of relief.

35

I Know Your Secret

Three weeks after the consultants met in Watford, Arthur Morgan was in Moorfields seeing Professor Khan.

Adil Patel had spoken to Holly Barron and reported nothing to help their problem. The eight consultants were still in the dark.

Arthur went through the usual batch of tests and examinations.

"Well, Mr Morgan, the eye is progressing very well," said Professor Mohamed Khan. "The transplant has settled in beautifully and your visual acuity is better than it has been for a very long time. I'm very pleased with the result. What do you think?"

"I am very pleased," said Arthur.

"Is there anything further before we send you home today?"

"Yes Professor."

"What is it?"

"The memories that I have of Sophie Coleman, are they permanent or will they vanish in time?"

"I don't think they are memories, Mr Morgan. It

is common for transplant patients to have dreams, hallucinations and even what they believe to be memories about their donor. We clinicians do not really know how these come about, but most often, they vanish over time."

"Most often, but not always?"

Mohamed Khan was becoming a little unsettled by this line of questioning. He and the other consultants had agreed to try to muddy the waters when asked by their patients about their memories of the donor. He had the distinct impression that the authoritative voice of Arthur Morgan knew more than he should. Mohamed was on his guard.

"It's just that I feel these memories getting stronger and more detailed. Was my donor a lady called Sophie Coleman?"

"What makes you think that transplant surgeons get the name of donors?"

"You must get some information so I thought you might be given a name, just in case the recipient wants to write to the family." There was almost a sigh of relief from Mohamed. He only wants to write a letter of thanks to the donor's family.

"If all you want to do is write to the family, we discussed this a long time ago. You write a letter and the transplant service forward it on. After that it's up to the donor's family to decide what to do next, if anything. There are cases where the organ recipient and the donor's family become friends. Most often the memory of a dead loved one is painful and the family do not want to be reminded. Much depends on the relationship between the donor and their family at the time of death. Also, there is the question of the cause of death. That can have an impact on how the donor's family react to a letter. When you write, be prepared for either scenario."

"Who would I write to Professor? Sophie was an only child, and her parents died in a horrendous accident some years ago. There is no one left."

"What makes you think that, Mr Morgan?" Mohamed Khan was astonished by this comment. He knew it was the truth, but how could his patient know?

"I have the memory of Sophie and I know that they are dead. Why else would she make Charlotte James, a friend, her next of kin?"

"This is preposterous. You are making this up." Mohamed was getting really worried. He was losing control.

"What about the other transplant patients who got one of her organs, do they get the same message you are giving me? Are they told that they are all making it up?"

You could have knocked Professor Mohamed Khan down with the proverbial feather as he heard these words from Arthur Morgan. He pulled himself together.

"What on earth makes you think that there are others? Have you had any more of these supposed memories that you think might be from the donor? Do you know something about her that you have not told us about yet?"

"No. I just know that there are others like me."

"How do you think you know this?"

"I got the information from my ears."

The Professor looked suitably stunned by this new piece of news. Random memories moving from a brain to an organ and then into another brain was one thing. Going to the brain through the ears was absurd. The Professor sputtered as he spoke.

"We did nothing that could have possible affected your ears. How do you think your ears came to possess this piece of fanciful information?"

"Eavesdropping, Professor. While I was getting ready in the ante room at my last appointment, I heard you talking to one of your colleagues about the others with this problem. From what I overheard, I deduced that I am not the only one as you have been at great pains to try to make me believe."

"Eavesdropping is rude. We might have been talking about something else completely." The Professor was trying to portray professional indignance. Arthur Morgan was not buying the charade.

"Is this memory thing related to one donor, or is it more widespread than that?"

"I've told you Mr Morgan that I cannot confirm anything, especially confidential medical information."

"Some time ago, you asked me to sign a document agreeing that my recollections of Sophie Coleman's memories could be shared. The story was that other specialists might have an interest in this phenomenon and be able help solve it. I suspect that those who had an interest were the doctors of the other patients who have this memory issue. I'm right, aren't I?"

"I cannot say."

"It is clear that you want to keep this a secret. I'd like to know why, particularly if you want my co-operation in the coverup."

"What do you mean by co-operation in a coverup?"

"It is plain to see that you have a reason for wanting this information hushed up. If I know why, I may be prepared to keep quiet."

"I have not told you that there are other transplant patients involved. If any other patients got a part of an organ donor's memory, don't you think the world might have heard about by now?"

"Maybe it's just this one donor. I'm sure if there had been others in the past it would have made the news. Since I remember nothing like it in the papers or on TV, I am guessing this is something new. Something you want kept under wraps. Something that worries you. If I know why, then perhaps I can be part of your conspiracy, if it's important."

"I don't know what you are getting at, Mr Morgan."

"Professor, I overheard you talking to one of your team. There are other patients who have the same dreams, hallucinations or memories as me. I thought I heard the number six or seven others. I am assuming that we are all recipients of an organ from the same donor, Sophie Coleman. I'm sure that I could find out if I really tried, but I am guessing you would rather I did not look."

"You are putting me in a very difficult position, Mr Morgan. If I tell you. what I know, you must promise never to reveal anything that I have said."

"I promise that if there is a very good reason for covering this up, I'll never reveal what you tell me."

"I'm doing this under duress."

"I understand, but I want to know."

There was a pause lasting several seconds before Mohamed Khan spoke.

"There are others who have this transplanted memory issue."

"How many? Two, three or more?"

"More. We know of seven others."

"Are they all connected to the memory of Sophie Coleman?"

"They are. We have found no other instance of this anywhere in the literature, anywhere in the world."

"Why is it such a problem?"

"Consider this, Mr Morgan. Before your transplant surgery, we had a long discussion about the benefits and problems of the surgery. All of the other surgeons will have had similar discussions with their patients. Can you imagine the anxiety of one of these patients who was being told that their doctors were going to remove the heart and put a new one in its place? This is nothing like a relatively simple cornea transplant."

"If your cornea transplant had gone wrong, you would have lost the sight in that eye. If the heart or lung transplant went wrong, the patient would die. The old heart or lungs are still working, all be it poorly, but they are still alive. Deciding to proceed with the operation is a very big decision."

"I can see that," said Arthur. "But why keep this memory thing a secret?"

"There are three reasons for secrecy for the moment at least."

"First. Suppose part of that conversation we had concerned the information that as well as the organ, you might receive a piece of the donor's memory, without any guarantee whether that memory be good or bad. If there is any possibility of this happening, medical ethics demand that we tell the patient. We would have to say that there was no certainty that you would get any of the donor's memory, just an educated assumption based on one isolated example."

"You might get nothing except the new organ. On the other hand, you might get a good or bad part of the donor's memory. You might get a memory of the happiest day of the donors life. On the other hand, you might discover that the donor was a murderer, or worse."

Professor Khan missed the look that crossed Arthur Morgan's face when he imparted this comment. Arthur brought his expression under control quickly.

"OK, Professor, that's one reason, and you said there were three. What are the others?"

"If this memory transplant was to become common knowledge, it might damage transplant surgery irrevocably. It could severely affect the patients with the new organs."

"How so?"

"You are considering registering for organ donation after your death. You can choose what organs can be taken, and which cannot. Suppose you chose to donate all organs but not your memory. We have no way of knowing how this occurs, and therefore we have no way of preventing it. We would have to reject your request to be a donor."

"Worse still, you might wish to donate only those memories that contain your religious or lifestyle beliefs. Since we have no way of selecting these memories, we cannot use your organs. The upshot of making this information public could be catastrophic for transplant surgery."

"Is there an answer to the problem?"

"Not at the moment. We do not know how this occurred. Because we do not have that information, we have no way of knowing who might pass on their memories as part of a transplant. We do not understand why each patient only got a small part of Sophie's memory and not all of it. None of the parts seem to overlap. We have no way of knowing whether the same section of every donors of memory will be transplanted with a particular organ. Is this an organ specific function? We don't know. Because we don't know how it occurred, we have no way of discovering a solution to block it."

"I now see the problem."

"Do you understand why we want to keep it secret. Can you imagine what the media would make of this if it became public knowledge? You, and the other patients would become celebrity freaks. You would have the media camped out on your doorstep trying to get an interview. That is the third reason."

"As a social worker, I see what that kind of celebrity does to people's mental health and wellbeing. I wouldn't wish that on anyone, especially me. I will keep the secret, Professor. Thanks for trusting me."

"Do you have any other reasons for asking this? Do you have a memory from Sophie that is perhaps sensitive?"

"I might have, Professor. It is something that might blow your desire to keep it secret wide open. In light of what you have told me, I need to take steps to ensure that does not happen. I need to work out how to act in the best interests of transplantation, the seven other patients and myself. There is a small chink in your armour that needs to be closed, permanently."

"What would that be?"

"Best I keep that to myself until I try to fix it. If I don't succeed, the whole world will soon know, and the manure will hit the fan. Wish me luck."

"Whatever you need to do, Mr Morgan, all us consultants wish you luck. Will you keep me informed?"

"I will. It might take me a few weeks, but I'll let you know.

36

An Unexpected Call

She was sitting in her living room watching television. The programme had no specific interest for her, but it was a noise that kept her company while her husband was out with his friends. The telephone rang, and she lifted the receiver.

"Hello."

"May I speak to Charlotte James please?"

"Speaking."

"Mrs James, my name is Arthur Morgan. You do not know me but I need to speak with you in private."

"Go away, Mr Morgan, if you're selling something, I don't want it."

"I can assure you Mrs James that I am selling nothing and I think we should meet face to face."

"Why would I want to do that? Are you some kind of pervert?"

"No. I can assure you that I am quite genuine."

"Why do you want to meet me?"

"I want to talk about your friend, Sophie Coleman."

"Sophie died last year."

"I know that."

"So, what do you need to speak to me about?"

"It's a very sensitive subject and best discussed face to face. I'll come to Tring to meet you in any venue that you choose, but not your house."

"How do you know that I live in Tring?"

"I know a great deal about you, Mrs James. That is one of the things that we should discuss."

"You said not my house. Why not?"

"Your husband might be there and I'm sure you would not want him hearing our discussion."

"Why?"

"Well, I want to talk about the events on the night of Friday thirty-first of October 2008."

"Oh." Arthur heard the shock in Charlotte's voice that was unmistakable.

"Oh, indeed Mrs James." Charlotte composed herself quickly.

"What do you know about that night that might interest me Mr Morgan?"

"Everything, Mrs James."

"Everything?"

"Yes, everything."

"Why would that interest me?"

"You know why, Mrs James. We need to talk face to face."

"Are you from the police?"

"No, Mrs James. I'm just an ordinary citizen. I am actually a social worker by profession."

"So, you want us to meet, and I get to choose the venue. Is that what you are saying?"

"It is."

"Somewhere public?"

"If that is your choice, but we should have enough

privacy for our discussion."

"OK. Given me a day or so to pick a place."

"I'll give you my number, but if I do not hear from you in a week, I will call again. It is imperative that we have this discussion. I'm happy to buy you a drink or a meal. Whatever you choose."

"Why?"

"All will become clear when we talk."

"I'll ring you. I don't want you ringing me when my husband is at home."

"Wise decision, Mrs James."

Two evenings later, Charlotte rang Arthur Morgan and arranged to meet in a small Italian restaurant adjacent to the Tring town car park on the high street.

"How will I know you?" She asked Arthur.

"Not to worry, I will recognise you. I'll see you next week."

37

Charlotte And Arthur Meet

Arthur entered Tring from the south and pulled into the marked public car park a few yards past the first roundabout. To the left of the entrance, he saw the sign for the Italian restaurant that Charlotte James had agreed to meet in.

He opened the door and scanned the few people sitting at tables. He recognised her immediately. She was seated in the corner by the window and looking onto the main street.

Arthur's memories of Charlotte were from Sophie, a female. He knew that Sophie admired Charlotte, perhaps even had a touch of Sapphic love towards her. Sophie had never acted upon her feelings, but the moment Charlotte James knocked on her door to tell her about Zac raping her, she had an immediate emotional connection.

As a male, Arthur was as impressed by her looks as Sophie was.

She had an oval face with a wide mouth and dark brown hair cascading down her shoulders. She wore little makeup, but what she wore complimented her looks. Over the table, Arthur could see she was wearing a white tee shirt, a striking yellow cardigan and a leopard-spotted scarf slung loosely around her neck. She had large gold-coloured hooped earrings.

She looked assured, but Arthur knew she would be anxious. He could see it in her eyes. Her strode over and offered her his hand.

"Mrs James, Arthur Morgan. I'm pleased to meet you."

"I cannot say the same, Mr Morgan," but she took his hand anyway. "Please sit and let's get this over with."

"I promised you a drink or a meal, perhaps we could order."

"Will it take that long?'

"That depends, Mrs James. There are some things that we need to resolve. At least one of those items will be to your advantage."

"All right, let's order."

The pair ordered their food, and Arthur suggested wine. "I'll only have a small glass as I am driving back to London."

As they waited for their food to arrive, Charlotte asked Arthur to explain why he wanted to meet.

"I need to start on 12 February 2012. As you know, the previous evening, your good friend Sophie Coleman had a crash on the M25. She died in the early hours of the morning of the twelfth."

"That same morning, I got a telephone call asking me to rush to Moorfields Hospital for a possible transplant operation."

"You know that Sophie was an organ donor because

you both signed up on the same day after her colleagues son died waiting for a transplant."

"How do you know that?"

"All in good time, Mrs James." At that point, their food arrived. Arthur had ordered Fegato di Pollo, a chicken dish, and Charlotte had a Wild Mushroom and Truffle Pizza. Arthur would have one small glass of Pinot Grigio, while Charlotte could consume the remainder of the bottle. Arthur thought she might need it. As they started their meals, Arthur continued.

"You asked how I know that you were both registered organ donors. It's because I was there when you did it."

"That's impossible, we did it online at Sophie's house."

"Yes, you did, and I was there. Let me explain, but it may seem a bit surreal. I'll go back to the night of 12 February. That night, I got a piece of your friend. If you look at my left eye, the cornea came from Sophie. The centre of my left eye once belonged to your friend."

"Oh my God." Charlotte spluttered and almost spilt her drink as she gasped in disbelief.

"Indeed. So, this small part of me was there when you or we both registered to be donors."

"Did I just hear you say we?"

"You did."

"When you say we, what do you mean? Where does Sophie fit into this we that you talk about?"

"By we, I mean you and Sophie, and since Sophie is now part of me, I also mean me."

"Are you claiming that since you have a small part of Sophie living in your eye, you are a sort of Sophie substitute?" Charlotte was an intelligent woman. She was having real trouble understanding what Arthur was telling her. She was bewildered and shocked by this statement.

"That is partly true," continued Arthur

"You must be on some sort of medication for your delusions. Either that, or you live in a parallel universe."

"I am on medication, but not the kind you think. I take drugs to stop my body from rejecting Sophie's cornea. I am not delusional, although it may sound as if I am."

"If you are not a nutter, how is it that you think you are Sophie's substitute?"

"This is the most surreal part of my story, and I swear to you that it is the truth. I'll tell you all about it if you are interested?"

"Why not? You invited me and are paying for the meal. I'll listen to your fiction."

"It is no fiction, as you will discover. Several months after my transplant operation, I started to have images in my head. It is difficult to describe how it first appeared, but I knew the donor of my cornea was female. Not long afterwards I knew that her name was Sophie Coleman."

"How could you know her identity. I thought the only way you could know was for you to write to the donors family and wait for them to reply. Sophie has no family left alive, so that can't be how you know."

"You are correct. No one told me, I just knew. It was information that was lodged in my memory."

"That's ridiculous. You got her cornea how could that have anything to do with your memory?"

"That's a very good question. The answer is that it is not really my memory. I believe that it is part of Sophie's memory that was transplanted into my body with her cornea."

"Did I just hear you say that the cornea that you got from Sophie came with part of her memory and that it is now in your head with your memories? That is bonkers."

"That's exactly what it sounds like when you say it like that but it is true, and it is there. Not only that, but more is to follow."

"What more?" Arthur could see the look of disbelief on Charlotte's face.

"Over the next few months I started to have more memories of a small part of Sophie's life. I explained what I was experiencing to the medical team at Moorfields, and I think they thought I was delusional. It seemed a perfect reasonable assumption to have."

"So, you do have some form of mental illness?"

"That might be the obvious conclusion, but it is not correct. It appears that somehow part of Sophie's memory was transplanted into me along with her cornea."

"Now that's just plain rubbish," said Charlotte. "How on earth could that have happened?"

"That's another good question, and one doctors are at a loss to explain."

"Did you say doctors, plural? Are there more like you?"

"I did, and yes there are."

"Your doctor can confirm this?"

"They can, but they will not."

"Why not? It's a great story. My paper would be delighted to run it."

"I'm sure they would, but they are not going to do that. This meeting is taking place to ensure that you do not leak the story."

"I could tell them anyway. It would not take them long to identify the patients. We could run your story as the man who broke the news."

"I'm sure all of that is possible, but it is not going to happen. After we leave this restaurant, you will forget about telling anyone."

"Why would I do that?"

"There are a number of reasons, chief among them being your freedom and your humanity."

"What do you mean my freedom and humanity?"

"Let me start with humanity," said Arthur reassuringly. "If you break this story then those of us afflicted by this memory transplant will become very reluctant media celebrities. Our quiet lives will be ruined. We all have enough trouble coping with the issues of our transplants. Coping with intrusive media attention might be devastating for some."

"Then there is the effect this story would have on future organ transplants. You are a registered organ donor. Would you like some part of your memory transferred to an unknown individual?"

"As you are aware, you can dictate the organs you want to donate and those that you do not. Should the register also include those memories you might want to donate, and those you do not."

"I never thought of that."

"You of all people would not want that to happen, would you?"

"No, absolutely not," gasped Charlotte. She was beginning to see the light.

"What about the people who are going to get the

new organ, what might worry them? Might they get nothing, a very pleasant memory or a really bad one? Might they get a memory of someone who was a murderer who was not yet caught?"

Arthur saw the shock on Charlotte's face. Deep in her mind, Charlotte did not believe this man could know her secret. It was impossible. She would brazen it out.

"Never the less, it is a great story that deserves to be told," said Charlotte.

"Does it really?"

"Of course, it does."

"Do you think people would believe it given that there are no previous known instances of this kind of anomaly anywhere in the world?"

"Why not. You have told me that there are several people who can confirm it. They can tell their stories."

"That means that I can tell mine. Is that what you want?"

"Why would I not want it?" Said Charlotte, trying to be nonchalant.

"Because I hold Sophie's dark secret, and it involves you."

"What are you talking about? I have no dark secret." Arthur could see from the look on Charlotte's face that she was unsettled by what he had said. She was beginning to believe that this man might know everything.

"Do you want me to spell it out for you?"

"Do your worst."

Arthur sipped his drink and looked Charlotte squarely in the eye.

"So be it."

38

Getting Away With Murder?

My part of Sophie's memory starts many months before her husband's death in what the police reported was an accident."

"Zac Coleman was a womaniser, a drunkard, a drug taker and violent towards Sophie, and you." He heard Charlotte gasp at this revelation.

"Yes, I know what you told Sophie about what he did to you. If I remember correctly, you turned up at Sophie's house on a Friday night, when you knew Zac would be out."

"How can you possibly know that?" Charlotte was very anxious when she heard this news.

"As I said Mrs James, I know everything that Sophie knew from the time you met her right up to the conclusion of the police enquiry. I have it all in here." Arthur tapped his head.

"You and she became very good friends. Actually, Sophie had a little crush on you. She also loved your children. She used to read them Beatrix Potter bedtime stories when she sat for you on your date nights with your husband."

"My God. I thought she might have had a crush.

We took our girls to the Lake District to see the Potter house because they loved Sophie's stories."

"Sophie took a number of beatings from Zac before she made up her mind what she was going to do. She told you that Zac had claimed to have killed a Mexican woman in the USA, didn't she?"

"She did." Charlotte's resolve was beginning to crumble.

"Sophie decided that she had taken all she was going to from Zac. She decided to kill him and make it look like an accident."

"Oh, no," gasped Charlotte.

Arthur looked around to ensure no one except her could hear what he was saying. He leaned in towards Charlotte.

"She talked to you about it and asked for your help. You objected at first, but after a particularly brutal beating she got from Zac, you agreed. You and she spent several months plotting his death."

"She siphoned some of his hoard of cocaine over many months. Little amounts that he would not notice. She made several trips to find a suitable spot for the final act. She planned to wait until he came home drunk one night. She provoked him into giving her a severe beating so that she would have the marks to show the police. She spiked his favourite Jack Daniels drink with the cocaine. She gave him just enough so that he would pass out. Then you and she, loaded him into his car. She drove him to the spot she had chosen, and you followed in your car."

"Then, you both sat him in the driving seat without fastening his seatbelt. You wore gloves to ensure that you left no fingerprints. Sophie also wore gloves so that her fingerprints did not sit over Zac's. The police would expect some of Sophie's prints to be in the family car."

"Sophie made sure that all of the seat and mirror setting were right for Zac, before she pointed the car at the edge of the steep tree lined gorge she had chosen. She set the car in drive and slammed the door shut as she jumped away. The car drove over the edge and smashed its way down through many trees and the odd rock before coming to a halt at the bottom. It was a long drop and the damage to the car was extensive enough to ensure he would not survive, as indeed he did not. How am I doing so far?"

"How would I know? I know nothing of this. I only turned up to comfort Sophie when she telephoned me to say Zac had beaten her again and she had chased him out of the house with a carving knife. I was still there the following morning when the police arrived to tell her the terrible news of her husband's death." Charlotte was desperate not to incriminate herself but knew the truth when she heard it.

"Indeed, you were. Sophie was grateful for your support, and for helping her convince the police of a completely different scenario. You went home after you drove Sophie back to her house. In this way the police could check your telephone call to confirm where you were and what you would tell them the following morning. They would see that both you and Sophie were at home when the call was made. It was a very good plan."

"None of this is true," said a tearful Charlotte.

"You know it is and Sophie was very grateful for all

your help in planning and carrying out this murder."

"What murder?" Said a very shaken Charlotte, who was trying hard not to cry. "There was no murder. The police ruled it an accident."

"Just as you and I planned, Charlie." The use of Sophie's pet name for her was the last straw in the collapse of Charlotte's crumbling resistance.

"What are you going to do?"

"That depends on you Mrs James."

"How so."

"If you attempt to break the story of memory being transplanted, I will be forced to tell my part of Sophie's memory."

"What can you say that would stop me?"

"You should have guessed by now that I know just how much help you gave Sophie. For example, how did a badly beaten Sophie get Zac into his car? He was a big heavy man. She needed the help that you gave. Then there is the question of how she got back from the crash site. That was in your car. You were Sophie's accomplice in the planning and execution of Zac's murder."

"No one will believe you."

"You are suggesting that people will believe your story about memory transplantation if you tell it. Why would they dismiss my story, especially as the other patients will have information about Sophie that can be verified?"

"How much do you want?"

"My price is very cheap and simple. I want you to say nothing about what I have told you."

"Why?"

"I have explained why. Telling this story might have

a devastating effect on transplant surgery around the world. Do you agree to my terms?"

"What if you die, suddenly?"

"Yes, I have considered that. Committing one murder is no worse than committing two. Doctors have been looking at every possible way in which this memory transplant could have happened. The patients who have received part of Sophie's memory have all told their part of Sophie's memory. The doctors are trying to work out whether there is something in her life that might have caused this. They have looked at the lives of the recipients and found nothing, so far. They have looked at any possible medical route that these memories can be lodged in an organ that will be transplanted. Again, they have found nothing. If they do not know how it occurred, they cannot stop it. The whole thing may be a freak of nature that will never be deciphered."

"In my part of this process, I have told the medics the facts surrounding Zac's death so that they coincide with your story and the police report. They have what you and Sophie wanted the police to know, not the truth as you and I know it."

"I have also recorded the true story, every last detail, including your participation. The meetings you had with Sophie before the murder with dates and times. The help you gave her to commit the crime. Your part in the coverup to ensure a police finding of accident. I also know that after the event, you spoke to Sophie about your feeling of guilt on a number of occasions. You were worried that the truth might get out. If you break this story, it will. My story is lodged with my solicitor who has it sealed with my will."

"If you try to break this story, I will publish that recording. What will happen to your life when that is published? What about your husband and two children?"

"You don't want money?"

"No. I want your silence."

"Why?"

"For the reasons I already gave you and because I have that part of Sophie's memory that shows what a pig Zac was. I'm no advocate of the death penalty but Zac is dead, and he got what he deserved. You keep our secret, and your part in his death is safe with me. I just wanted you to be reassured that you need not worry. That is the only reason that I wanted us to talk. I knew you worked for a news outlet and the temptation to break a story might get the better of you. I hope that I have explained the consequences if you do. What do you say Charlie?"

"What can I say? You have me over a barrel."

"I don't look at it like that. I want you to know that if this story is broken by you, then I will break you in return. If the story happens to escape by another route, then your secret is safe with me. I will relate my part of Sophie's memory just as you, she and the police reported. Is that agreed?"

"Yes. I was worried that I might get discovered, even if Sophie is dead. Are you the only one with this information?"

"As far as I know, I am. If any of the other patients had it, I would expect the consultants to be more anxious than they already are. The prospect of any of the others having the information does not bear thinking about. I believe you are safe."

"There is one other thing I would like to know."

"What would that be?"

"Did you and the girls scatter my ashes in the Lake District as promised?"

"Oh my God. You know about that?"

"I do. As I said earlier, Sophie used to read Beatrix Potter books to your girls when she babysat for you and Paul. The girls loved the books. Did you keep Sophie's collection of Potter books for your daughters?"

"I did."

"Is Sophie at rest in the Lake District?"

"She is. We took them there during summer break. The girls and I scattered them on the Potter land. Down a rabbit hole. It seemed appropriate."

"That's nice. Sophie felt sure that you would do as she asked. She or I will keep your secret if you keep ours."

"I will. Thank you."

"You are welcome. Shall I get the bill?"

39

All Clear?

After his conversation with Charlotte James, Arthur called Moorfields to speak to Professor Khan.

"Mr Morgan. Did you manage to resolve the chink you thought we had in our story?"

"Yes, Professor, I believe I have."

"What was the chink?"

"There is another individual who knows of this memory issue, and she works for a newspaper."

"Oh my God," gasped Mohamed.

"An excellent sentiment. She would have been anxious to tell the story but I believe that I have convinced her to stay silent."

"How did you manage that?"

"I explained the damage the story could do to transplant surgery worldwide. She is a registered donor. She understands the medical difficulty of perhaps having an unknown part of the donor's memory transplanted with the organ."

"I also explained how my life would be affected if this story surfaced. She was beginning to come around to my way of thinking when I hit her with the clincher."

"What was that?"

"Best you do not know Professor, but it involved blackmail, of a sort."

"Blackmail. What information do you have that gives you that sort of leverage?"

"As I said, it is probably best you do not know. I warned her that if she ruined transplant surgery, and my life by releasing this information, then I have the information that would ruin her life, job and marriage."

"You were correct. It is best I do not know what you have. I cannot be a party to blackmail."

"Are you and your colleagues any closer to understanding how this happened?"

"No. It seems the harder we look, the less we find. We have a meeting in two week to consider a way forward. At the moment, we are all in the dark."

"Good luck, Professor."

"Thank you, Mr Morgan. I'll see you at your next appointment."

40

Cut Worms Forgive The Scalpel

As Arthur Morgan talked with Professor Khan, Graham Bradbury was reading articles that interested and alarmed him. It was a paper on the stability of memory during brain remodelling.

He came across these articles when he was trying to finalise his thoughts on memory transfer from one human to another during an organ transplant. His basic assumption was that it was impossible until he read the papers.

There was research on worms, Artic Ground Squirrels, wasps, weevils, mushrooms and other species. The most interesting was the work on worms (Planaria).

The researchers trained worms using classical conditioning protocols.

The worms were trained using conventional methods. The researchers cut them in half and allowed them to regenerate.

After regeneration, the half of the worm that contained its brain retained its training in repeat tests.

Interestingly, the half that lost the brain required less training to learn when compared to untrained animals, suggesting some residual learning or memory.

Even more interesting was what happened afterwards. The enlarged survey involved cannibalism.

Researchers fed trained worms to untrained worms. The worm cannibals were trained alongside unfed worms to compare the two groups. Fed worms learned more quickly than those who were unfed.

This threw up the possible existence of an engram that might transfer into untrained animals. The researchers postulated that RNA was the active agent in this memory transfer. Worms injected with RNA extract from trained worms demonstrated this memory retention while RNAse treatment abolished it. Graham knew RNAse was a treatment that could remove RNA from genomic RNA samples. This experiment strongly suggested that RNA might be a vehicle for memory transfer. RNAse treatments seemed to offer a method of shutting it off.

Other experiments suggested that stem cells might be the cellular location for memory. There was even a suggestion in some of the papers he read that showed memory in some mammals can be transferred offline and relocated by the acquired organ to the storage region of the brain.

As he read these articles, Graham noted that they all suggested further research, meaning there was no proof - yet.

Much more research would be needed to prove or disprove these reports. Nevertheless, Graham was troubled by what he had read. These were eminent people in their fields producing work that could have significant implications for organ transplantation. A cursory look at these data suggests memory might be transplanted offline by transplanting a new organ.

Was there a remote possibility that RNA or stem cells could replant memory? If so, why had the problem with eight patients not been seen earlier? The more he thought about it, the more he worried.

Perhaps it had been seen before, he thought. Maybe it had been suppressed, as he and his colleagues were trying to do. The so-called dreams of patients that medical teams scorned as mental complications might be transferred memories.

While he was dismayed by this thought, Graham also thought there could be beneficial effects in several medical issues, not least Alzheimer's. What the hell was he going to do now?

Two days later, he approached Allan Cunningham. He had printed out all the papers he had read and had a summary of their findings attached.

"Shit," said Allan Cunningham when he read Graham's summary. "These were animal experiments, not human trials."

"Correct. There is nothing to suggest that human experiments would produce different results."

"I'll read the papers before our next Zoom meeting with the others. This is good work Graham. Alarming, but good."

The Zoom meeting started as arranged. Allan had emailed each consultant with the papers and Graham's summary.

"Ladies and gentlemen, thank you for your attendance. Have you all received and read the papers I sent?"

They all had.

"What Graham's paper has shown is that it might just be possible for donor memory be transferred offline during organ transplantation. Work is being undertaken to look at the role of RNA or stem cells. There may be other methods that are being worked on, but not yet published. Over to you, Graham."

"I was at first dismayed to discover that RNA might have a role in this memory transfer, if indeed it does occur. Given our experiences with these eight patients, this may be a reality. This is worrying for all the reasons we have previously discussed."

"The answer to this particular issue lies in one experiment. RNAse switched of the learned memory. Doing that to the organ at harvesting, might be the answer."

"As for stem cells, we are all aware of the work that is going on to produce replacement organs using this field. One of the issues faced is switching off the antigens in the stem cell to kill the patient's immune response."

"It may be that we will have the ability to switch off these antigens in the replacement organ, before we have wholly artificially produced organs. We might also have the ability to switch of memory transfer, if it happens, using RNAse."

"What I have difficulty believing is that memory transfer might always have been a factor in our patient's recovery when we accused them of having psychological issues and nightmares."

Keith Isaac raised his hand to speak.

"I have to report that my liver transplant patient is dead. He was found in his tiny flat with a carving knife in his heart. The coroner had no option but to pass an open verdict. There was no forced entry to his flat. His fingerprints were the only ones on the knife. His blood alcohol level was off the scale, he had drunk a whole litre bottle of neat vodka. On the balance of probability, he committed suicide, but there was no note left. Without specific evidence that he wanted to end his life, the coroner could only use the test for criminal behaviour."

" My concern for the death of this man has much to do with the memories he got from Sophie Coleman. He was a man who was violent towards his wife and children, and they left him. His memories of Sophie were of the beatings that she had received from her husband. He complained about it at appointments, and I had him see our psychologist frequently. It was not just the memory of the beatings she received, but he also stated he could feel the blows. My patient having the memories was difficult enough to believe. His claim to actually feel the beatings was impossible to comprehend."

"My best guess would be that he got first-hand experience of the beatings that he had dished out to his wife, and could not stand it."

"I believe that the memories he had from his donor might have caused him to take his own life. If, as Graham has suggested, there is a mechanism for memory to be transplanted, then we must do what we can to aid the research into switching it off."

Mohamed Khan spoke next. "My patient overheard me talking to my senior registrar about this issue. He cajoled me into telling him that he was not the only one. I did this because he said that he knew of a flaw in our plan to keep this secret. He did not elaborate, but after I told him why we wanted this kept secret he agreed that we were correct. He offered to plug the loophole."

"How could he do that?" Asked Isobel Mason.

"He did not elaborate, but telephoned me to say we had the all clear. I asked how he had done it and he said that he appealed to her sense of fair play and he threatened her with blackmail. At this point I did not want to know what he held over this person. I suspect that he has a memory from Sophie Coleman that he has not shared with us. A memory that sacred this person into agreeing to keep our secret. For that reason, and Keith's patient, I think we should do all we can to resolve this possible problem for us and every other transplant surgeon."

"What can we do?" Asked Adil Patel.

"Perhaps we speak to those people who are doing research into memory," said Giuseppe Da Silva. "Some of the people that Graham unearthed might be looking for human trials. We take so many samples from our patients, perhaps some might be useful to researchers."

"That's not a terrible idea," said Nicola Stephens. "Our ultimate aim must be to get to the seat of this problem. Helping those researchers might help us. I suggest we speak to some of them and see what happens. They may say no."

"Ladies and Gentlemen," said Allan Cunningham. "It seems that memory transplant might just be possible. It may even be that we have seen it in the past and not recognised it. Our staff have commented on occasion as to how our patients personalities have changed since their transplant. This may be another manifestation of what we are now witnessing."

"If the dreams and nightmares we have seen in previous patients were indeed transplanted memory, then our present predicament is more worrying to me. These dreams or nightmares have been woolly, vague, imprecise or nebulous. Our eight patients memories are anything but. They are explicit, lucid, distinct and sharp. I come back to something that was suggested at a previous meeting. The clarity of recall of our eight patients is unique. Perhaps, as was suggested, there was something unique about our donor that we have not discovered."

There was general agreement on this point, and the meeting slowly slid into a highly technical and scientific discussion. There was a discussion about how synapses work, neurotransmitter chemicals, body memory, memory during brain remodelling, and any number of our highly technical sub-subjects before Allan Cunningham called a halt.

"Clearly there is much to ponder. However, we are all highly skilled, very experienced surgeons, but we are discussing the frontier of experimental medical research in a field in which it seems none of us are experts. I suggest we return to the suggestions that we volunteer to be involved with expert researchers in this field. It is in all our interests, given what Keith and Mohamed have said about their patients."

"This is early in the process of discovery. I think we would all want to ensure, that whatever the outcome, it is beneficial to our patients and our area of surgical expertise."

There were murmurs of consent.

"I will task Graham with contacting some researchers in this field of offline memory transfer, to see if there is anything that we can contribute, such as samples or histories. Do you agree?"

They did, and the meeting closed with a promise from Graham to keep them informed if anything new arose.

41

Life After A Death

The death of Sophie Coleman was not on the same scale as the way Christians view the death of Jesus. Still, she died so that people might live.

The eight recipients of her organs all had a variety of outcomes. Some of them were good, some not so good.

Jack Gibson died as a result of an eight-inch kitchen knife plunged into his heart. The coroner's inquest was obliged to bring in an open verdict as she had to apply the criminal standard for evidence. Had Jack died a year later, the coroner might have been able to use the less strict standard based on the balance of probabilities. In that case, the coroner might have recorded Jack's death as suicide.

Jack left no note. As a result, why he might have committed suicide would remain a mystery. Jack's wife, Amy, believed that his death was no considerable loss to humanity. Keith Isaac might be inclined to agree, but he would never say so.

The small bowel that Megan Thomas received was perhaps the most likely organ to fail of its own accord.

Intestinal transplants are notoriously unpredictable because the organ is not sterile. So, it proved for Megan. Hoping to save the organ, Kasara Doshi increased Megan's antibiotic and immunosuppressant cover. He and his team did their best to save the organ, and Megan, but she died in 2017, just after her fiftieth birthday.

Jacob Marshall met a young woman during one of the support meetings run by a Heart Charity. She was a member of the charity team. She piqued his interest because he felt he had met her previously. He had difficulty pinpointing where or when.

The young woman was named Carlotta Jiménez. She had an oval face, a wide mouth and dark brown hair cascading over her shoulders. It took Jacob three meetings to pluck up his courage to talk to her.

"Have we met before?"

"Yes. You have been to two of my previous meetings."

"That's not what I meant," spluttered Jacob. "I meant had we met before you started these meetings?"

"I do not think so unless you lived in Manchester?"

"I've never been there. I have this picture in my head of a woman that I met somewhere. The image I see is exactly like you."

"Sorry, I can't help you." She turned and walked away.

It took Jacob two more meetings to approach her again.

"This is going to sound strange, but I think I know where I saw you before."

"How can it be strange?"

"When I had my transplant, I was under anaesthetic when I had a very vivid dream that included meeting a woman. The woman I met could be your twin."

"Did you just say that you met me in a dream you had under anaesthetic. Strange is not the word. Are you perhaps delusional?" She was laughing.

"Perhaps I am, but I remember the dream because it was so clear and vivid in my head. If that means I am delusional, then that is what I am."

"Well Jacob, that is the strangest chat-up line I have ever heard."

"I didn't really mean it as a chat up line, but since you mention it, would you like to have dinner with me somewhere nice?"

"Was a dinner invitation part of your dream?"

"No, and yes. Going out somewhere with the beautiful lady was."

"A beautiful lady? In that case, I can hardly refuse. I'm interested to find out more about your dream, and how I got into it in the first place."

Jacob could not help but wonder how the image of this woman had got into his head. He would never know that Carlotta Jiménez could be mistaken for the twin sister of Charlotte James. It may have been Sophie's love for Charlotte that triggered her likeness in his dream.

They made arrangements for their meal, and as a result, their relationship slowly blossomed.

The consultants had discussed how it took around four months for Sophie's memories to become apparent. It was another part of the mystery. They might have forgotten that electrical impulses in the human nervous system travel very rapidly. If the human brain says to walk, it does not take the legs several months to obey.

Encouraged by his memories of her time at university, Richard Ainsworth thought he might try further education. He wanted to improve his employment prospects.

After making the decision, his first thought was to check out the courses at De Montford University. He had never been there, but he knew the layout of the university and city from Sophie's memories. When he looked at the entry qualifications to some courses that interested him, he realised he was not eligible.

Richard turned his attention to the Open University. He chose the Certificate of Higher Education in Computing & IT with a second module in Business Studies. It required about eighteen hours a week of study for two years. The fees were very reasonable compared to full-time education at a university. The principal part of the course was computing and IT. There were four possible choices for a second module. Richard chose business as the others held no interest to him.

What also drew him to distance learning at the OU was that he could continue to work and earn as he studied. He was also able to stay close to his medical team. Not having to change hospital consultants was very reassuring.

It came as no surprise that when he qualified, Richard got himself a job in the tourism industry. It was where Sophie was employed.

Arthur Morgan's eyesight recovered very well after his transplant. Within seven months, his eyesight had become as good as it would get. He still wore glasses rather than contacts. He felt they made him look more intellectual and managerial. It suited his promotion to management in his Local Authority. He had been wearing glasses for many years and did not feel adequately dressed without them.

His promotion was a double-edged sword. Every member of his team had caseloads that were far more than what any sane person would consider safe

In 2018, he attended a budget meeting in his authority. The politicians informed him that cuts in his department staff numbers were needed. Arthur argued strongly against them. This was a financial decision made by politicians. It had no regard for the human consequences.

He was to lose two members of staff. He argued that if these cuts went through, he might lose others. They were at breaking point, and most suffered anxiety attacks and depression brought about by their inability to do their jobs properly. He asserted that the increased workload would inevitably mean something serious would be overlooked or missed. He argued the money saved from these cuts might have other financial consequences, such as legal costs. His argument received no sympathy from his political masters.

As he predicted, one of his ablest workers, Ms Farina Zafar, resigned in tears.

"I can't cope anymore, Arthur," she said. "I go home at night and cry most of the time. I can't give our clients the time that they need and deserve. It's like pushing water up a hill. I quit."

"I understand Farina," sympathised Arthur. "We will miss you."

"I will not be in tomorrow, Arthur. Here is a sick note from my GP. It will cover the whole of my notice period. She wants me to have rest and to see a therapist to try to help me. I'm sorry to leave you in the lurch, but I just can't do it anymore."

Four months after Farina left, his department suffered a serious case review. What he had claimed at the budget meeting had come to pass. A young woman had died. It was an avoidable death with an adequate complement of staff. The local authority managers sought to blame their social work staff to divert blame away from their cost-cutting responsibilities. It would take Arthur only three months to decide that he had enough. He resigned to take a job in the police criminology department.

If he thought his troubles were behind him, he was about to discover they were not. He returned home one evening to find his wife agitated. She had discovered that their daughter, Chloe, aged thirteen, had discovered boys. The discovery was made worse by some social media posts that young Chloe had written. Amanda thought that some of what she had posted was inappropriate. When he saw the posts, Arthur agreed.

"Where is she?" Asked an angry Arthur.

"Upstairs in her room, crying. I went mad at her when I saw what she has done. I might have gone over the top," said Amanda.

"From what I see here, I doubt it. Does she understand what she has done?"

"I'm not sure she does. She stormed off shouting at me that old people don't understand the young. This, according to her, is normal for her friends. She hates us, she says."

"I'm sure she does at the moment. We need to devise a strategy to convince her that this is wrong and could have serious repercussions."

Arthur and Amanda sat for over an hour while they discussed what to do. Arthur had been used to holding difficult and sensitive conversations with his clients. This was very different because it was his daughter, his firstborn. He was not sure his training had prepared him for this.

The first meeting with his daughter and wife did not go well. Chloe was very defensive. "It's what everyone else is doing," she shouted.

"You're not everyone else, Chloe. You're our daughter, and this opens you up to any number of dangers. I have seen far too many young girls like you pass through my old social work department. I have no desire to see that happen to you."

The fraught conversation continued for about half an hour. Arthur looked at his wife and gave Chloe an ultimatum.

"If you do not mend your ways, I'll take your mobile away from you and disconnect your laptop from the internet."

"I need it for my homework," she shouted.

"You can use the free computers at the library for that. If you show that you are responsible, I will let you have them back. You have one week and I'll be monitoring you very closely."

Chloe did rein back on her social media activities. However, she knew her mother and father were watching her like a hawk. She was extra careful about what she posted.

When a boy attacked one of Chloe's friends at school, her attitude started to change. She had been one of Chloe's close friends on social media.

The student councillors at the school targeted Chloe and the raped girl's other social media friends. Chloe reluctantly admitted that her parents might have been right in reminding her of the dangers. Arthur and Amanda heaved a sigh of relief for now.

Abigail struggled with her new lungs. She was considerably better than before the operation, but she kept having setbacks. In early 2022, Abigail was having difficulty with her breathing. She was breathless and wheezing. Adil Patel decided that she needed a stent inserted into her narrowing airway.

The procedure went well, but before she could go home, Abigail felt very unwell. It was almost as if she had influenza. Abigail was kept in the hospital to await the results of her blood tests.

These tests revealed that she had contracted the new virus named COVID-19. It was a respiratory disease, and for someone with transplanted lungs and a weak immune system, it was dangerous.

There had been news of this new disease sweeping worldwide from its origins in Wuhan, China. It was a time when the country needed a competent leader. Britain had Boris Johnson, maybe the most incompetent Prime Minister since the war. If you need proof, here is a quote from Dominic Cummings, Johnson's chief advisor. When asked why he would not let Johnson do an interview on the BBC with the formidable presenter, Andrew Neil, he replied. Why the fuck would we put a gaffe machine, clueless about policy and government, up to be grilled for ages by Andrew Neil? Enough said.

Abigail's situation deteriorated. It got to the point where she needed to be in ICU. Abigail had been in the unit after her operation. What she encountered now was bedlam. The unit kept expanding into any available space because of the number of seriously ill patients arriving daily. Gone was the one-to-one nursing. It was clear that overstretched staff were buckling under the strain. Many were scared. Patients were dying at an alarming rate. Eight days after being admitted to ICU, Abigail died on a ventilator.

During the week ending 24 April, when Abigail died, the England & Wales NHS would expect to see about ten and a half thousand deaths a year. The Office for National Statistics reported twenty-two thousand deaths that week. Eight thousand two hundred were from COVID-19. Just over five weeks earlier, they reported only five COVID-19 deaths.

Rose Prentice made it to December 2020 before she also died as a result of contracting COVID-19.

Rose slipped and fell in her home at the beginning of December that year. She fractured her left ulna and had a severe blow to her temple on the left side. She was knocked unconscious by the fall. She was taken to hospital when an ambulance became available.

In the A&E department, they thought that she might have a significant concussion. They admitted her to a side room in a general ward. She got the side room because she was on immunosuppressant drugs and prone to infection.

This was during the second wave of COVID-19, and all hospitals were under enormous pressure. Staff were working criminal hours using inadequate protective equipment bought from hastily approved suppliers, many of whom had no experience making it. However, the suppliers were friends of the government party ministers. They got fast-tracked approval with minimal checks on their competence.

Despite the hospital staff's best efforts, Rose caught COVID-19 in hospital. She was transferred to the ICU a week later. Placed on a ventilator, Rose was visited by the flip team every four hours. Their job was to turn Rose on to her front or back to give her a better chance of survival. Like many before, it did not succeed. Rose died on Christmas Eve.

Rose's cremation in January 2021 was a sad affair, and not just because she had died. There were government restrictions on the number of people who could travel or attend such events.

The young Simone Jessop had a very successful recovery from her operation. Apart from the memories, Simone might also have inherited a cytomegalovirus (CMV) infection from Sophie. CMV is an infection that may also have lain dormant in Simone and become active after the transplant as a result of her weakened immune system due to anti-rejection medications. Antiviral drugs controlled the problem.

Early in her recovery, Simone thought she might

like to become a nurse. Gracie Wilson had been a significant influence on her. The more she thought about it, the more she became turned off by nursing as a career. Hospitals were full of people with infections. Nurses were always one of the first points of contact for infected patients. Transplant patients had to be wary of any infections. Nursing was not the career for her with her reduced immune system.

After searching for an answer to her career problem, Simone decided she wanted to be involved with IT App development. Her computer skills were good but not up to the standard required to be a programmer. The other route was to be a copywriter. She felt that with the correct training, this was a job she would enjoy. She remembered using the mySugr app and how it could be better. If she could do something like that, she would be happy.

Having studied what skills were needed, Sophie took courses in copywriting, marketing and design.

She eventually became a junior lifestyle copywriter for a new business, selling a new way for people to manage their money. The fact that her father had a background in the financial sector in the city did help. He could see what the company was trying to achieve and offered Simone suggestions that she could take to her employer.

Simone was good at her job. She was also a source of ideas for her employer. She became recognised as someone who would succeed and scale the promotion ladder.

What of Graham Bradbury? He had the prospect of

one day becoming a Consultant Surgeon. It was what he had been aiming for.

He was becoming increasingly tired of the politics and competence of senior management in the NHS. As a consultant, he could do more lucrative private work in his free time. The NHS regularly advises well-to-do patients on long waiting lists to look at private treatment. Some in his department's waiting list had done that. They had jumped the queue to see Mr Cunningham at the local private hospital.

Graham was not sure he liked the practice of queue jumping. He could understand the frustration of his patients and the financial motive of his consultant.

As the eight departments continued to seek an answer to their issue, Graham continued reading everything he could related to memory transplantation.

There was a wealth of information on the internet and in medical journals. Articles in the medical journals were properly constructed scientific works. It was easy to follow their methodology and tests that led to the conclusions or findings.

Information on the internet was much more reckless. There was a lot of it, and most of it anecdotal. There was minimal, if any, scientific rigour attached to many of these articles.

Many of these articles were along the same lines. Joe Soap had a kidney transplant. He also got the memory of information leading to the murder conviction of a previously unknown suspect. It was amazing what people would ascribe to organ transplantation without any proof.

Graham had an old uncle who was in care after being diagnosed with Alzheimer's disease. The more he read, the more he wondered if there was any significant evidence that RNA could transplant memory. If it could, there might be a method of storing a person's RNA for future transplantation, should the worst happen. There were sperm and egg banks for fertility treatment. Why not an RNA memory bank for Dementia or Alzheimer's?

Far-fetched perhaps, but weirder ideas had become useable.

When a new opportunity arose, Graham left his job at St George's Tooting. He joined the research department of a private pharmaceutical company doing experimental work with memory and Alzheimer's treatments. The work was rewarding clinically and financially. It was a good career move.

Authors Note

When I started this story, I imagined it as a work of fiction. I wrote it as such. Why would I think otherwise? I had read nothing to the contrary and my life around all things medical suggested it would be fiction.

I was born on the grounds of a hospital where my parents worked. When I left school, I had a short stint in a research laboratory, long stints in the Armed Forces medical branch and a multi-national pharmaceutical company. My final post was in an international medical charity. When I retired, I became a volunteer for two medical charities. I have familial experience with the ups and downs of a kidney transplant.

This is a hotchpotch of experiences with no specific knowledge of the story I wanted to write.

The story needed to have factual information about organ transplantation. To that end, I researched the subject using a variety of sources. The NHS Blood and Transplant department was the most valuable source of that information. Their excellent website (nhsbt.nhs.uk) has a detailed description of every aspect of each type of organ transplant. Each organ has a complete description of the process. The site starts with the question, 'Is a transplant right for you'?

There are sections on the risks and benefits of receiving a transplant, what happens at the transplant centre, and ends with a long section on living with a transplant. The only organ not listed on their site is a cornea transplant. Fortunately, there is equally good information about the pros and cons of this procedure from Moorfields Hospital and others.

Patient stories of their experiences of transplant surgery proliferate on the internet. Many medical charities use those stories to aid understanding and generate income.

I have familial experience with one type of transplant. It is nowhere near enough to create stories for eight individuals of varying ages who need and receive different organs.

I read patient stories from all over the world, some uplifting, some not so good. My familial experience falls into this latter category. The stories I created for my patients are an amalgam of everything I read and experienced in my family. If any of the eight stories mirrors that of any person, living or dead, it is pure coincidence.

Throughout history, doctors have made many attempts to transplant organs. By about 800 BC, Indian doctors had probably begun to graft skin.

In the 16th Century, the Italian surgeon Gasparo Tagliacozzi discovered what later became recognised as transplant rejection. He found that skin from a different donor usually caused his procedures to fail.

In 1818, Dr James Blundell invented the instrument for transfusing blood from one human to another.

In 1905, an Austrian doctor called Eduard Zirm conducted the first cornea transplant. It saved a man's eyesight. It was the world's first successful organ transplant.

In 1954, the first successful live donor kidney transplant took place in the USA, with the recipient receiving the organ from his twin brother.

Most older people will know that in 1967, the South African surgeon Christian Barnard replaced the diseased heart of a dentist. The man died eighteen days later, but the floodgates were now open.

Today, organ transplantation is a common medical procedure.

After researching organ transplantation, I shifted my focus to memory. I was curious about how humans make memories and where they are stored. There is extensive global research on memory, and I learned that while there is significant knowledge, there are still many unknowns.

Initially, my research centred on human memory, but I couldn't find any connections that would make memory transplantation feasible.

Imagine my shock when I stumbled across a couple of papers showing the possibility of memory transplantation in worms (planarians).

These papers suggested that my original premise that memory transplantation was fictional might be wrong. Memory transplants might already be occurring without being recognised as such.

I recently met some nurses in my local shopping centre. They were distributing information for a diabetes charity. I told them about my story. They had been part of transplant teams and had seen the personality of organ recipients change over time. Based on personal observation, they wondered if these changes were brought about by receiving part of the donor's memory. It was only a thought, they said.

So, is memory transfer possible? The research on planarians suggests that even after head-chopping and regeneration, it shows that somehow planarians retained, regrew or enhanced their memory. It was suggested that the body retained memory and not only the brain. This was a conclusion based on observation rather than empirical proof.

Most scientists believe that only the brain stores memory, but some brainless animals seem to manage to learn and remember things. The research in planarians suggests that RNA plays a part in these animals. There is no published evidence that it does in humans. In the absence of supportable evidence, the answer is currently not known. I do not doubt that some researchers are trying to rectify this lack of evidence.

As I write this story, I am left to conclude that however our bodies store memory, it is very complicated.

Within the story, I raised the ethical question of memory transplantation. People in the UK can choose which organs they might want to donate and which they do not. If memory transfer is possible, how does the world deal with that situation?

It's not hard to imagine an autocratic ruler dictating that only cultural or religious memory supporting them can be transplanted. The Nazis in World War Two experimented with this on unwilling subjects.

It raises the question: just because we can, should we? History tells us we will, especially if it brings financial benefit or increased power.

www.ingramcontent.com/pod-product-compliance
Lightning Source LLC
Chambersburg PA
CBHW020630220526
45464CB00001B/92